The Institute of Biology's
Studies in Biology

Antibiotics and Antimicrobial Action

Stephen M. Hammond
Ph.D.
Department of Biological Sciences, Purdue University,
Indiana, USA

Peter A. Lambert
Ph.D.
Microbiology Division, Glaxo Research Ltd, Greenford,
Middlesex

Edward Arnold

© S. M. Hammond and P. A. Lambert 1978

First published 1978
by Edward Arnold (Publishers) Limited
25 Hill Street, London W1X 8LL

Board edition ISBN: 0 7131 2683 3
Paper edition ISBN: 0 7131 2684 1

All Rights Reserved. No part of this publication
may be reproduced, stored in a retrieval system,
or transmitted, in any form or by any means, electronic
mechanical, photocopying, recording or otherwise, without
the prior permission of Edward Arnold (Publishers) Limited.

Printed in Great Britain by
The Camelot Press Ltd, Southampton

General Preface to the Series

It is no longer possible for one textbook to cover the whole field of Biology and to remain sufficiently up to date. At the same time teachers and students at school, college or university need to keep abreast of recent trends and know where the significant developments are taking place.

To meet the need for this progressive approach the Institute of Biology has for some years sponsored this series of booklets dealing with subjects specially selected by a panel of editors. The enthusiastic acceptance of the series by teachers and students at school, college and university shows the usefulness of the books in providing a clear and up-to-date coverage of topics, particularly in areas of research and changing views.

Among features of the series are the attention given to methods, the inclusion of a selected list of books for further reading and, wherever possible, suggestions for practical work.

Readers' comments will be welcomed by the author or the Education Officer of the Institute.

1978

The Institute of Biology,
41 Queens Gate,
London, SW7 5HU

Preface

A major contribution of science to the well-being of mankind has been the isolation of a variety of natural products from microorganisms and the development of their use into the techniques of modern chemotherapy. It is difficult for anyone born in the developed world in the last forty years to comprehend the ravages of terminal septicemia, pneumonia, tuberculosis, bubonic plague, cholera or typhoid. These names have lost the fearsome ring they had at the beginning of the present century and are no longer certain death sentences. They have become the names of medical conditions which will respond to the appropriate therapy. A mere handful of compounds, the antibiotics, with the aid of certain synthetic agents, have brought these diseases under control. Almost every reader of this book will have taken an antibiotic preparation at some time in his life, yet few are aware of the variety of mechanisms by which antimicrobial agents interfere with the cellular function of invading pathogens without affecting the host. Nor are they conscious of the years of intensive research that are required for the discovery and development of new antibiotics. In this small volume we hope to give the reader some insight into the advances made in the understanding of antibiotic activity and to demonstrate that the important discoveries which have been made in the uses of antibiotics are not restricted to medicine. Study of the mechanism of action of antibiotics has revealed much about the biochemistry and physiology of all living cells.

1978

S. M. H.
P. A. L.

Contents

Preface		iii
1	**Antibiotics as Chemotherapeutic Agents**	1
	1.1 The origins of chemotherapy 1.2 Antibiotics and selective toxicity 1.3 Antibiotics and microbial growth 1.4 Antibiotics as secondary metabolites 1.5 Antibiotic-producing micro-organisms 1.6 Molecular mechanisms of antibiotic action	
2	**The Study of Antibiotic Activity**	8
	2.1 Isolation of new antibiotic-producing strains 2.2 Determination of minimum inhibitory concentrations (MIC) 2.3 Diffusion assay methods 2.4 Antagonism testing (cross diffusion test) 2.5 Synergy testing (the paper strip gradient test)	
3	**Inhibitors of Cell Wall Synthesis**	20
	3.1 The bacterial cell wall 3.2 Antibiotics affecting the cell wall 3.3 Penicillins 3.4 Cephalosporins 3.5 Cycloserine 3.6 Bacitracin 3.7 Vancomycin and ristocetin	
4	**Membrane-active Antimicrobial Agents**	28
	4.1 Structure and function of cell membranes 4.2 Membrane-active antiseptics 4.3 Cyclic polypeptide antibiotics 4.4 Ionophore antibiotics 4.5 Gramicidin A 4.6 Polyene antifungal antibiotics	
5	**Inhibitors of Nucleic Acids**	37
	5.1 Nucleic acids 5.2 Structure of DNA and RNA 5.3 Biological roles of DNA and RNA 5.4 Semi-conservative replication of DNA 5.5 Transcription and RNA biosynthesis 5.6 Antibiotics which affect nucleic acid function 5.7 Agents which interfere with nucleotide biosynthesis 5.8 Agents which interfere with the polymerization of nucleotides by impairing the template function of DNA 5.9 Agents which interfere with enzymes involved in nucleic acid synthesis	
6	**Inhibitors of Ribosome Function**	45
	6.1 Protein synthesis, the function of the ribosome 6.2 Streptomycin 6.3 Tetracyclines 6.4 Chloramphenicol	
7	**Metabolic Inhibitors**	50
	7.1 Introduction 7.2 Interference with folic acid 7.3 Sulphonamides 7.4 Other compounds affecting folic acid metabolism	
8	**Prospects for Chemotherapy**	53
	8.1 Antibiotic resistance in microorganisms 8.2 Biochemical nature of antibiotic-resistance 8.3 The spread and control of antibiotic-resistance 8.4 Development of new antimicrobial agents 8.5 Antibiotics as tools in scientific research 8.6 The future	
Further Reading		64

1 Antibiotics as Chemotherapeutic Agents

1.1 The origins of chemotherapy

Chemotherapy, the use of chemical agents to damage invading microorganisms without damage to the host, has a history as long as the science of microbiology. It is perhaps appropriate that Louis Pasteur, the father of microbiology, made the first recorded observation of antibiotics, i.e. the inhibition of one microorganism by products diffusing from another. In 1877, Pasteur and Joubert showed that anthrax bacilli were killed when the culture became contaminated by certain other bacteria. Therapeutic applications of this observation rapidly followed. Living bacterial and fungal cultures or their extracts were applied directly to wounds with variable results. An extract of the bacterium *Pseudomonas aeruginosa*, known as pyocyanase, was marketed for over fifty years and its use only ceased with the advent of the sulphonamides. In 1917 Greig-Smith reported that several actinomycetes (a group of filamentous bacteria common in soils) produced products with antibacterial activity. So began the systematic search of products of microbial metabolism for possible antimicrobial agents, which has continued to the present day. In 1937 Welsch isolated actinomycetin from a species of actinomycete belonging to the genus *Streptomyces*. This compound showed marked antibacterial properties, but proved too toxic for therapeutic use. Dubos (1939) isolated tyrothricin (later shown to be a mixture of gramicidin and tyrocidin) from the bacterium *Bacillus brevis*, which was capable of curing systemic bacterial infections in mice. The mixture proved poisonous when used systemically in man but could be used topically (application directly to the infected area) as for example in throat lozenges.

The first therapeutically useful antimicrobial agents were not antibiotics. It was the dream of the great German chemist Paul Ehrlich (1854–1915) to synthesize 'magic bullets', organic chemicals capable of killing pathogenic organisms exclusively. To this end he devoted his life and many of the concepts of modern chemotherapy descend directly from his thoughts. His laboratory synthesized many hundreds of organic compounds of arsenic (arsenical preparations had been used previously for protozoan infections but their use was limited by the acute toxicity of arsenic to man) and examined their antimicrobial spectrum and toxicity to mammals. In 1912 Ehrlich and Bertheim produced salvarsan (compound number 606) which proved effective against the causative organism of syphilis with negligible toxic side effects. Chemical modification of the arsenical molecule produced drugs which were more soluble and more selective in their action (neosalvarsan). Ehrlich also examined the property of dyes to react with specific tissues and hoped to find dyes both specific for and lethal to pathogenic organisms. This approach was continued by Domagk who in 1935 examined a series of azo dyes for possible antimicrobial properties. He found that a compound called prontosil, first synthesized three years previously, showed antimicrobial activity against systemic staphylococcal infections in man. Chemists began a series of modifications of the prontosil molecule to reduce toxic side effects while retaining or enhancing the antimicrobial spectrum and produced a series of compounds, the sulphonamides. These could be taken orally, and rapidly found use in the treatment of pneumonia, puerperal fever, meningitis, gonorrhoea and staphylococcal infections.

A significant advance in chemotherapy was made in 1940 when Florey and Chain solved the problem of extracting the labile antibacterial substance from *Penicillium notatum* and named penicillin by Alexander Fleming in 1929. Florey and Chain examined the activity of penicillin against a range of bacteria *in vitro*, determined its mammalian toxicity and used penicillin to treat experimental infections in mice, with remarkable results. The problems of large scale penicillin production were solved in the United States and purified penicillin was used to save many lives in the Second World War (Fig. 1–1). In 1944 Waksman isolated streptomycin from a species of *Streptomyces* isolated from soil. Since then continued searches have revealed many thousands of antibiotics.

1.2 Antibiotics and selective toxicity

Antibiotics may be defined as substances produced by microorganisms antagonistic to the growth or life of others at high dilution (but excluding organic acids, peroxides and alcohols produced by many microorganisms). The name antibiotic is in part a misnomer. There is little evidence to suggest that organisms synthesize antibiotics to fight natural enemies. In soil, antibiotic-producing strains of actinomycetes are in a minority; the level of antibiotics existing in soil being negligible. Similarly penicillin production by *Penicillium notatum* is rare in nature. If antibiotic synthesis gave an advantage, antibiotic-producing strains would predominate in nature and free antibiotics would be detectable in the environment. As antibiotic production occurs late in the microbial growth cycle, this would not help in natural competition. The true function of antibiotics in nature is debatable.

Although the term antibiotic is useful, the distinction it makes between naturally occurring and synthetic antimicrobial agents is not of fundamental importance, e.g. chloramphenicol was originally a natural product but is now produced by direct chemical synthesis. Through trivial usage the term antibiotic has become synonymous with chemo-

Fig. 1–1 This poster was produced during the Second World War to boost the production of penicillin which was needed for the casualties expected in the D-day landings.

therapeutic agent, i.e. chemicals that can directly interfere with the proliferation of microorganisms at concentrations that are tolerated by the host. Their essential feature is selective toxicity, a feature which distinguishes chemotherapeutic agents from disinfectants and antiseptics. Disinfectants, e.g. bleaches, chlorxylenol (Dettol) and cresols (Lysol) are strongly bactericidal but are completely non-selective, i.e. are poisonous to mammals. They are generally used to remove bacteria from inanimate material. The difference between disinfectants and antiseptics is only one of degree. Antiseptics, e.g. quaternary ammonium compounds (Cetrimide) and chlorinated phenols (TCP) are less irritant to tissues and may be applied topically. Antiseptics are nevertheless poisonous when taken internally. Many of the antibiotics isolated have proved to be far too toxic for use in man and therefore fail to become chemotherapeutic agents.

The selective toxicity of chemotherapeutic agents is achieved by exploiting intrinsic differences between the host and invading organism. For example, the agent may act upon structures present in the invader but absent in the host. Penicillin halts microbial growth by preventing the synthesis of the bacterial cell wall to which there is no comparable structure in mammalian cells. Biochemical pathways of microorganisms and man often differ and these points of difference may be exploited. In bacteria sulphonamides inhibit folic acid synthesis which is necessary for growth; man has no need to synthesize folic acid, taking sufficient from his diet, so that his metabolism is unaffected by sulphonamides.

Bacteria can be divided into two groups on the basis of their affinity for a simple chemical stain, the Gram stain. Organisms that take up the stain and are not decolorized by alcohol are termed Gram-positive, while those that can be decolorized are Gram-negative. The action of antibiotics is generally against a restricted range of pathogenic bacteria, be they Gram-positive or Gram-negative. Those antibiotics which are effective against a variety of pathogens (both Gram-positive and Gram-negative) are termed broad spectrum antibiotics.

1.3 Antibiotics and microbial growth

When bacteria are inoculated into a growth medium, there is an initial lag phase before the onset of growth (Fig. 1–2). This is followed by a phase of exponential growth, the logarithmic phase. As available nutrients decline and toxic products accumulate, growth slows and then halts, i.e. the stationary phase.

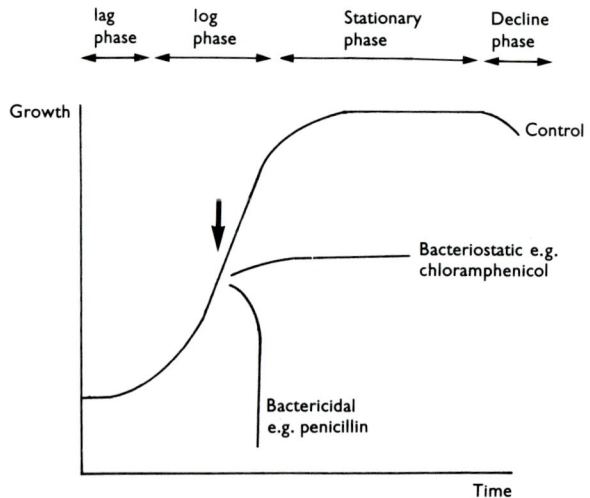

Fig. 1–2 Effect of antimicrobial agents upon the growth of *Escherichia coli*. Antibiotic added at time shown by arrow to exponentially-growing culture.

Antibiotics have been divided into two groups: bactericidal antibiotics (e.g. penicillin) which kill or cause the lysis (dissolution) of invading bacteria, and bacteriostatic antibiotics (e.g. chloramphenicol) which merely inhibit bacterial growth and replication (Fig. 1–2). Bacteriostatic agents rely on the host defences finally to defeat the infection and if antibiotic therapy is suspended microbial growth may resume. This division, however, is not clear cut, as bacteriostatic agents may become bactericidal if the concentration is raised and vice versa.

1.4 Antibiotics as secondary metabolites

Antibiotics are part of the vast range of natural products known as secondary metabolites. Primary metabolism is the interrelated series of enzyme-catalysed reactions which provide living cells with energy, with synthetic intermediates and with key macromolecules such as proteins and nucleic acids. Primary metabolism is essentially the same for all living species and, as a result of finely balanced metabolic control systems, its intermediates, the primary metabolites, rarely accumulate. Secondary metabolism is characteristic of lower life forms and secondary metabolites are often species specific. In contrast with primary metabolites, from which they are created, secondary metabolites often accumulate in substantial quantities and are often excreted into the medium. This is particularly true of the secondary metabolites of microorganisms cultivated in aqueous medium. Secondary metabolites tend to be formed late in the growth phase of microorganisms and their function may be to protect the microbial cell from adverse conditions, i.e. secondary metabolism is a means of detoxification. Secondary metabolites may also inhibit macromolecular synthesis preventing abortive growth under hostile conditions.

The production of an antibiotic is usually confined to a single group of closely related organisms. This rule of specificity derives from the position of antibiotics as secondary metabolites. A major exception to this rule is cephalosporin N which was first isolated from the filamentous fungus *Cephalosporium* but was later isolated from *Streptomyces*. The rule of specificity also applies to whole classes of subunits used to build up antibiotics, e.g. the amino sugars found in the polyene antifungal antibiotics produced by *Streptomyces* are not found in other antibiotics. The distinction between primary and secondary metabolism is not always well defined. Primary metabolites may accumulate in non-physiological quantities, for example *Aspergillus niger* may excrete the primary metabolite citric acid and certain bacteria produce excess nucleic acids. The antibiotic nisin has been considered by some workers to be a primary metabolite.

1.5 Antibiotic-producing microorganisms

The taxonomic distribution of antibiotic-producing organisms is restricted to relatively few groups. This may not, however, represent the real situation in nature, only reflecting the areas of search.

The capacity of true bacteria to produce antibiotics is limited as regards chemical structure, number (approximately 5% of described antibiotics are from true bacteria) and producing species. All antibiotics of bacterial origin used in medicine are produced by the genus *Bacillus*, all are polypeptides and include the polymyxins, gramicidin and tyrocidin. Although possessing similar chemical structures these antibiotics differ in their mechanisms of action; polymyxins are active against Gram-negative bacteria and only weakly active against Gram-positive organisms. The reverse is true for gramicidin and tyrocidin.

Fungi are a more important antibiotic-producing group (approximately 20% of described antibiotics derive from fungi). Chemotherapeutically useful antibiotics have been found only among the order *Aspergillales* (a group of filamentous spore-forming moulds) and include the penicillins, the cephalosporins and fusidic acid.

The actinomycete genera *Streptomyces* and to a lesser extent *Nocardia* are easily the most important antibiotic-producing group (approximately 75% of described antibiotics). Actinomycetes synthesize a wide range of antimicrobial compounds with diverse chemical structures, many different mechanisms of action and varied antimicrobial spectrum and include amphotericin B, chloramphenicol, kanamycin, erythromycin, rifamycin, streptomycin, neomycin, novobiocin, tetracyclines and vancomycin.

1.6 Molecular mechanisms of antibiotic action

Although all antibiotics are able to prevent the growth and multiplication of susceptible bacteria, microbial inhibition by different antibiotics is not brought about by a common route. This is not surprising as the term antibiotic brings together a diverse group of chemical compounds with little in common excepting antimicrobial activity. Antibiotics bring about microbial inhibition by interacting with specific cellular components and disordering cell metabolism. Over the last twenty years the mechanisms by which specific antibiotics interact with components of microbial cells have been extensively studied. It became evident that antibiotics could be grouped according to the site in the cell they attacked. The five major lines of attack are shown diagrammatically in Fig. 1–3 and will be discussed in more detail in Chapters 3 to 7.

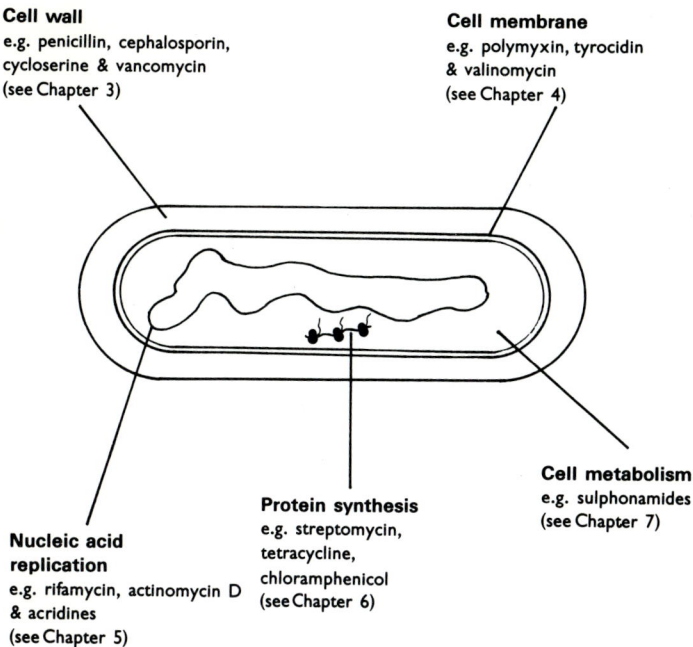

Fig. 1–3 Summary of the biochemical targets for drug action against a generalized bacterial cell.

2 The Study of Antibiotic Activity

2.1 Isolation of new antibiotic-producing strains

The techniques used in the search for new antibiotics are often direct adaptations of Fleming's initial observation of the diffusion of penicillin from *Penicillium notatum* growing on an agar plate. Strains of microorganisms are isolated from natural environments and pure cultures examined for the ability to prevent the growth of pathogenic organisms.

Soil is the natural habitat for actinomycetes and filamentous fungi and so it is not surprising that examination of soils has provided a wealth of antibiotic-producing microorganisms. The diversity of microbial life in soil depends upon many factors, including the physical and chemical properties of the soil, vegetation cover and climate. For example, cultivated soils from temperate climates contain a large number of microorganisms but of restricted variety, whereas uncultivated soils from tropical climates usually contain a smaller total number but a much greater variety. Tropical and uncultivated soils are therefore more productive of antibiotic-producing species. Antibiotic-producing microorganisms are not restricted to soils; useful strains have been isolated from habitats as diverse as sewage, rotten fruit, hospital isolates and the sea. Any new microbial isolate may have the potential to produce antibiotics.

The examination of isolates for antibiotic production is termed antibiotic screening. The soil or other potential source may contain many millions and several species of microorganisms per gram. The first step is to separate the microorganisms and to obtain pure cultures. A few grams of the original material is diluted in sterile distilled water, transferred to a sterile petri dish (a shallow glass or plastic dish with a lid) containing solidified nutrient media and distributed over the media surface with a sterile spreader (Fig. 2-1). Microbiologists use many different types of nutrient solidified by addition of a substance called agar. Growing microorganisms on the surface of solidified nutrient agar confines the added organism to one site. As the organism multiplies it forms a visible colony containing many thousands of cells all of which are descendants of the original microorganism. Microorganisms often have specific growth requirements. To realize the full microbial flora from a natural environment, samples are placed on a range of nutrient media and grown at different temperatures and in the presence or absence of oxygen. After incubation for 2–7 days several colony types will be seen on the plate. A small portion of each colony is removed with a sterile bacteriological

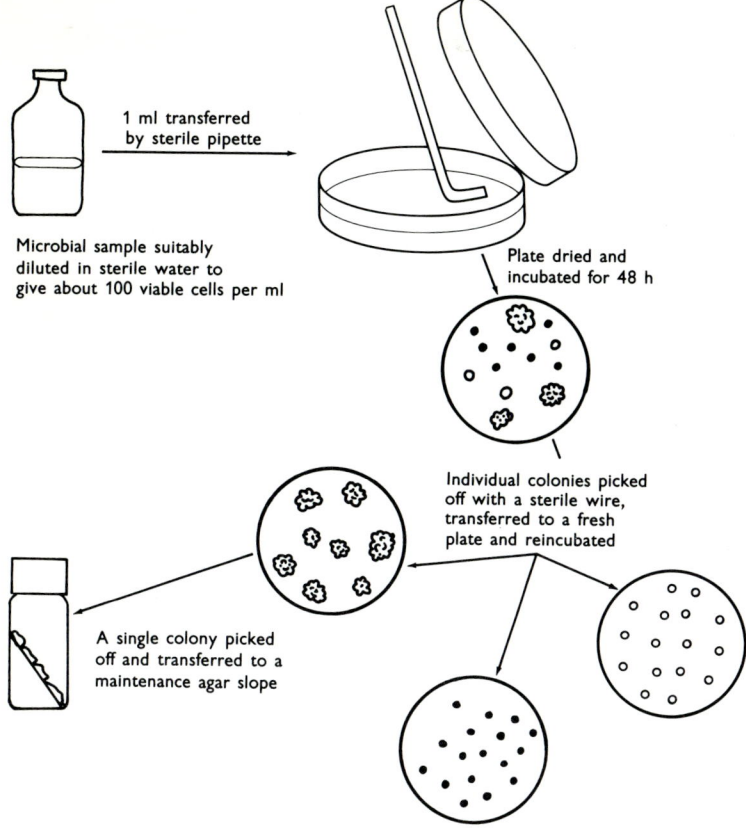

Fig. 2-1 Isolation of pure microbial cultures from natural environments.

loop and replated. After further incubation the now pure culture can be transferred to an agar slope in a small screw-top (or universal) bottle. Slope cultures can be stored for several months.

A single soil sample may contain a hundred different organisms and each must be treated separately. To determine whether any of the isolates produce compounds with antimicrobial activity each organism is spread over the surface of a nutrient agar plate and incubated until growth is confluent (covers the entire plate surface). Discs are then cut out of the plate with a sterile cork-borer (No. 4) and transferred aseptically to a plate seeded with a test organism (Fig. 2-2). The test organism is usually a mild pathogen with moderate resistance to known antibiotics, e.g. *Staphylococcus aureus* or *Escherichia coli*. Seeded plates are prepared by adding a suspension of the pathogenic bacteria to molten nutrient agar at 45° C (a higher temperature will kill the bacteria and at lower

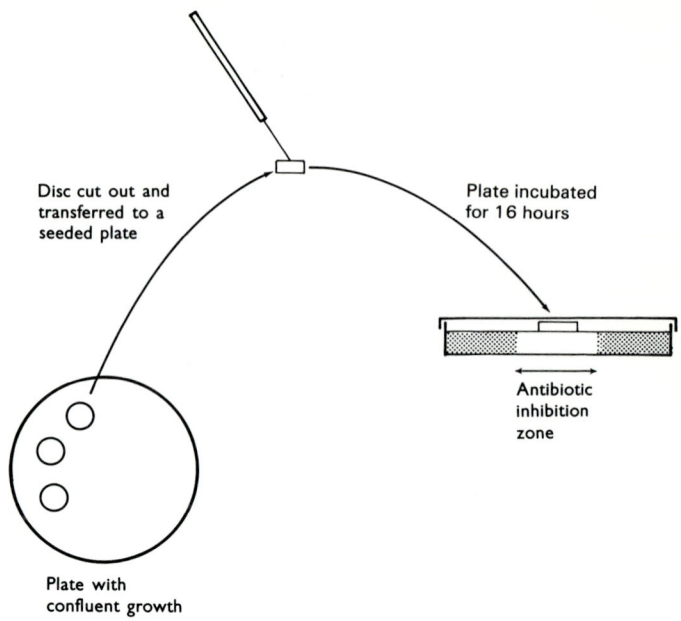

Fig. 2–2 Demonstration of antibiotic production (see also Fig. 1–1).

temperatures the agar will set), mixing thoroughly, pouring the mixture into a petri dish and allowing the agar to set.

The plate, with implanted disc, is incubated for 16 hours at 37°C. If the isolate is producing antibacterial substances they will diffuse from the disc and prevent the growth of the test organism. This results in a clear zone of inhibition, which contrasts with other regions of the plate where microbial growth, within the agar, makes it opaque (Fig. 2–3).

If the isolated strain has useful antimicrobial properties the next step is the production of the antibiotic in sufficient quantity for further investigation. It is important to find the conditions (e.g. temperature, pH, aeration, medium composition, etc.) for optimum microbial growth and antibiotic production.

There may be several different antibiotics in a filtrate from a single isolate. A bioautograph can be used to decide whether the isolate is producing one or several antimicrobial compounds. Spots of concentrated culture filtrate are placed on a sheet of chromatography paper and placed in a chromatography tank (Fig. 2–4). Antimicrobial compounds present in the culture filtrate diffuse with the chosen solvent at different rates. When the solvent front has reached the top of the paper (10–16 hours) the paper is removed and dried. The sheet is then placed on the surface of a nutrient agar plate seeded with a suitable test organism.

§ 2.1

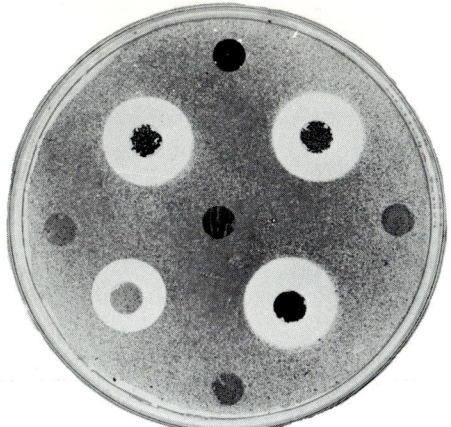

Fig. 2–3 The industrial isolation of new antibiotic-producing microbial strains. Plugs cut from isolate cultures are placed on a plate seeded with *Staphylococcus aureus* and incubated. Inhibition zones appear around those cultures producing antimicrobial compounds.

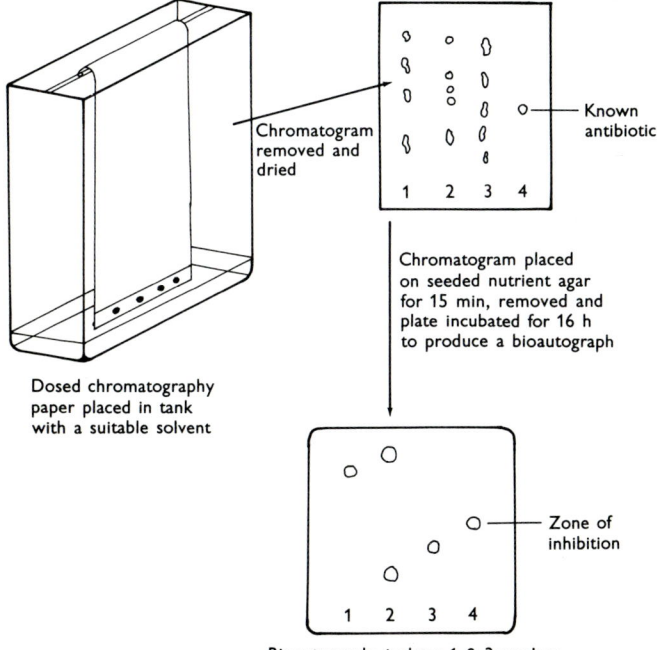

Fig. 2–4 Separation of antibiotics from culture filtrates by chromatography and the preparation of a bioautograph.

After 15–30 minutes the paper is peeled off and the plate incubated overnight. Clear zones of inhibition can then be seen corresponding to the position of the antibiotic on the chromatogram.

The antibiotic must next be purified and subjected to as many analytical techniques as possible to confirm that the compound is unique. If the chemical structure can be determined the antibiotic may fall into one of the existing groups and this knowledge will aid the evaluation of its properties.

There is an established procedure, in most countries, to which all new drugs must be subjected before they can be released for general medical use. Of the many new antibiotics isolated from nature every year less than one in a thousand is authorized for human use and less than one in ten thousand has novel properties or advantages over the antibiotics in current therapeutic use.

2.2 Determination of minimum inhibitory concentrations (MIC)

It is of great importance that the relative resistance of microorganisms to each antibiotic is known for there is no point in giving a patient an antibiotic that has no effect on the invading pathogen. As an index of antimicrobial activity microbiologists and clinicians use the minimum inhibitory concentration (MIC), defined as the lowest antibiotic concentration that will inhibit the growth of a specific organism. Small bottles of liquid growth media, containing graded doses of antibiotic, are inoculated with the test organism. After suitable incubation growth will occur in those bottles where the antibiotic concentration is below the inhibitory level and the culture will become turbid (cloudy) from the large number of microorganisms present. Growth will not occur above the inhibitory level and the bottle will remain clear (Fig. 2–5). To confirm the result samples can be taken from the bottles and plated out onto nutrient agar. The plate containing the lowest antibiotic concentration which shows no microbial colonies after incubation is the one with the MIC.

2.3 Diffusion assay methods

Diffusion assays are based on the ability of antibiotics to diffuse from a confined source through a nutrient agar gel and create a concentration gradient. If the agar is seeded or streaked with a sensitive organism, a zone of inhibition will result where the antibiotic concentration exceeds the MIC for that particular organism.

Ideally the antibiotic sensitivity of a pathogen isolate should be determined before antimicrobial therapy begins. The gradient of antibiotic concentration can be easily made by placing a sterile large square plate at an angle and adding antibiotic-containing agar (Fig. 2–6). When the agar has set, the plate is placed on a level surface and an equal

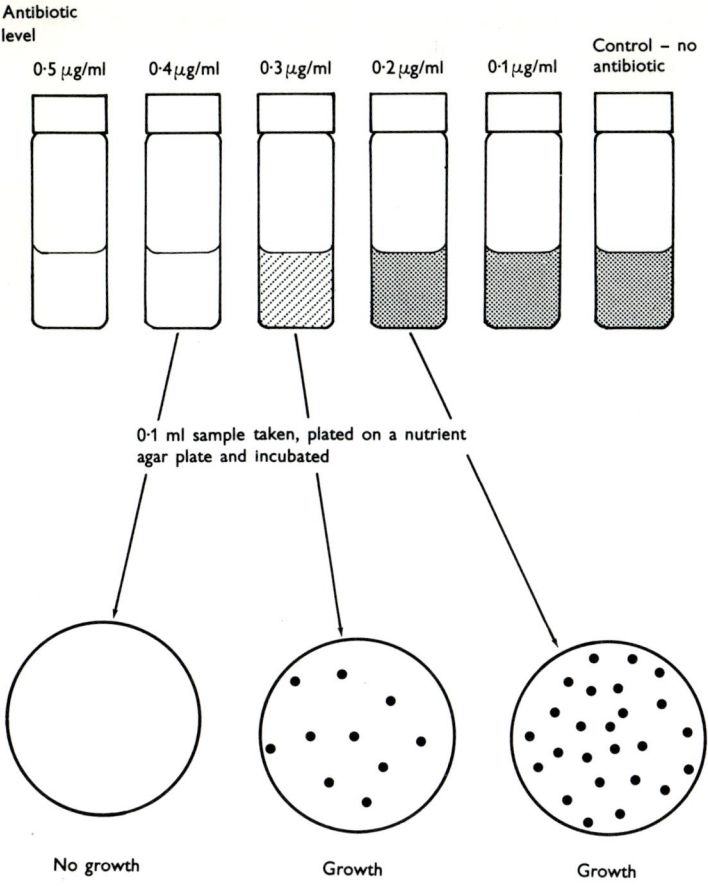

Fig. 2–5 Determination of minimum inhibitory concentration (MIC). Bottles viewed two days after inoculation. Lowest antibiotic concentration preventing microbial growth (MIC) = 0.4 µg/ml.

volume of nutrient agar without antibiotic is added. When set and allowed to dry, the plate is streaked with isolated pathogens and a test strain of known antibiotic sensitivity. The streak is made to extend across the plate surface from the region of low to high antibiotic concentration. After incubation the length of streak that has grown can be measured and compared with the streak-length of the known organism.

For rapid screening of antibiotic sensitivity predosed discs of filter paper are commercially available. Particularly useful are Multo-disks®(Oxoid Ltd., Basingstoke), which have a number of projecting arms the tips of which are impregnated with a range of antibiotics. A

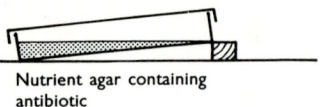
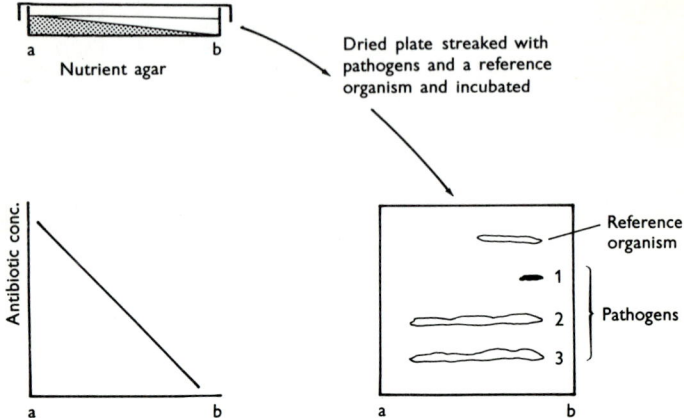

Fig. 2–6 The gradient plate technique. Organisms 2 and 3 are more resistant and organism 1 is more sensitive than the reference organism.

Multodisk is placed on a nutrient agar plate seeded with a pathogenic isolate. After incubation clear zones of inhibition will remain around the tips containing those antibiotics to which the isolate is sensitive (Fig. 2–7). Antibiotics diffuse through agar gels at different rates, e.g. penicillin G and chloramphenicol diffuse rapidly while polymyxin diffuses slowly, so that zone sizes produced when testing specific isolates with different antibiotics are not directly comparable. It is wrong to assume that an organism is more susceptible to one agent than to another on the basis of a larger zone size.

Antibiotics exert their antimicrobial action at very low concentration levels that are difficult to detect by physical or chemical methods. Diffusion assays can be used to determine accurately very low antibiotic concentrations. Provided that certain conditions are fulfilled, the diameter of the zone of inhibition around a confined antibiotic source is proportional to the logarithm of the antibiotic concentration. These conditions are:

(a) The composition of the nutrient medium must be standardized. As some components of laboratory media affect certain antibiotics, specially formulated antibiotic assay media have been developed. The depth of

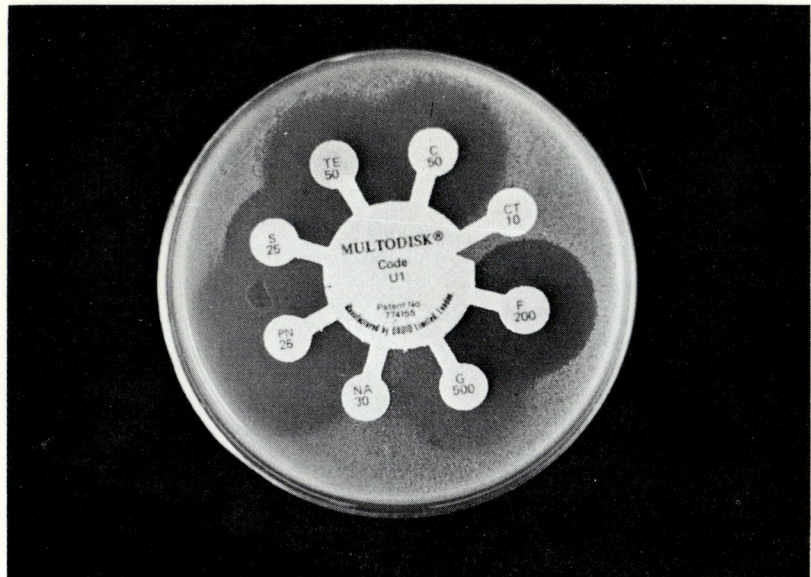

Fig. 2–7 Use of a 'Multodisk' to determine the antibiotic sensitivity of a pathogenic isolate. Familiar antibiotics on this disc are chloramphenicol (C), ampicillin (PN), streptomycin (S) and tetracycline (TE), the other drugs are nitrofurantoin (F), sulphafurazole (G), nalidixic acid (NA) and colistin sulphate (CT) a polymyxin.

medium and the degree of hydration of the agar gel must also be standardized.

(b) The inoculum must be of constant size, growth rate and antibiotic sensitivity. If the inoculum is too large the zones of inhibition might not be clear. Antibiotic assays are carried out with specific microbial strains of known antibiotic sensitivity (obtainable from microbial reference collections).

(c) The incubation time and temperature must be constant. These two factors affect both antibiotic diffusion and microbial growth. If the incubation period is too long, regrowth may occur into the inhibition zone.

(d) For accurate results the experiment must be repeated several times and the zone diameters produced by known antibiotic concentrations compared statistically with inhibition zones elicited by the test solutions.

These conditions are incorporated into the technique commonly used to determine antibiotic concentrations, the large plate assay. 200 ml of tempered antibiotic assay medium, seeded with the specific antibiotic test microorganism, are aseptically poured onto a levelled large square (25 cm^2) assay plate. After cooling and drying the plate surface, six rows of equidistant wells are cut in the agar with a sterile cork borer (No. 4). The plate is then laid over a latin square (a random distribution of the numbers 1 to 6, set out in 6 rows of 6). (Fig. 2–8.)

2	3	6	4	1	5
6	1	5	3	4	2
3	2	4	6	5	1
4	5	3	1	2	6
1	4	2	5	6	3
5	6	1	2	3	4

Fig. 2–8 A latin square used for 6 × 6 antibiotic assay.

In the assay for streptomycin the test organism *Staphylococcus aureus* is added to streptomycin/yeast extract agar. The series of wells labelled 1 and 2 are used for two streptomycin standards (4.0 μg/ml and 0.5 μg/ml). Wells numbered 3 to 6 can then be used to assay unknown streptomycin concentrations (test solutions must be adjusted until they lie in the range of the assay, i.e. 0.5 to 4.0 μg/ml). After overnight incubation inhibition zones will develop (Fig. 2–9). Figure 2–10 shows a typical calibration curve prepared from the average results of wells 1 and 2. The zone

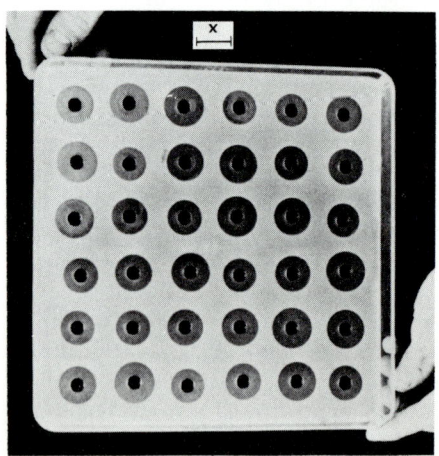

Fig. 2–9 An antibiotic assay plate. The wells are dosed with different antibiotic concentrations according to a latin square and incubated. The diameter of each inhibition zone (x) is measured.

§ 2.4

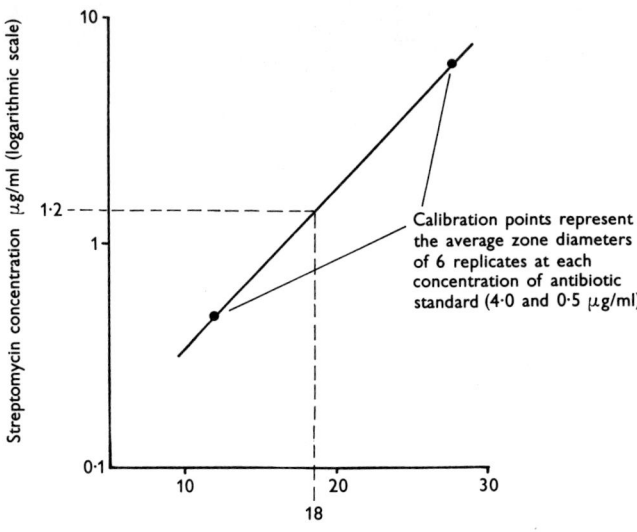

Fig. 2–10 A typical large plate antibiotic assay calibration curve. Assay for streptomycin using *Staphylococcus aureus* as test organism (plate incubated for 18 hours at 37°C).

diameter increases linearly with the logarithm of the antibiotic concentration in the range 0.5 to 10 µg streptomycin/ml. The streptomycin concentration in the test solutions can be determined from the calibration curve, e.g. if a 1 in 100 dilution of one of the unknowns has an average zone diameter of 18 mm, the test solution has a streptomycin concentration of 1.2 × 100 µg streptomycin/ml.

2.4 Antagonism testing (cross diffusion test)

It is often useful to be able to test for substances suspected of interfering with the action of an antibiotic (antagonism). Two strips of filter paper, one soaked with the antibiotic, the other soaked with the suspected antagonist, are placed so that they cross at right angles on a petri dish containing nutrient agar seeded with a microorganism sensitive to the antibiotic (Fig. 2–11). If there is no antagonism the antibiotic inhibition zone will be linear (A) but if the added agent is antagonistic a diagonal borderline (B) will result (representing the constant inhibitor: antagonist ratio that just permits growth).

2.5 Synergy testing (the paper strip gradient test)

Antimicrobial synergy is said to occur where the antimicrobial effect exhibited by two compounds, when used together, is greater than the sum

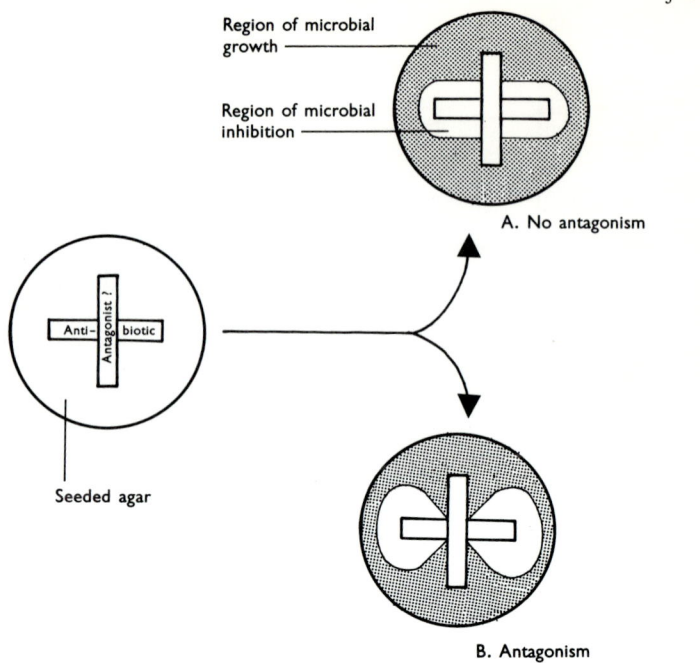

Fig. 2–11 The cross diffusion test.

of their actions when equivalent amounts of the agents are used separately. The synergistic agent might be another antibiotic, but might also be a substance that itself has no antimicrobial properties.

The cross diffusion test is too insensitive to test for synergy. This is because synergistic effects are relatively small (a five to tenfold synergistic effect is considered remarkable) and difficult to detect by diffusion assay (due to the logarithmic relationship between inhibition zone diameter and antibiotic concentration). The paper strip gradient test is used to test for synergy. An antibiotic gradient plate is prepared (see section 2.3) with the antibiotic concentration chosen so that growth of the test organism is inhibited at about halfway. The overlaying agar is seeded with a sensitive organism. A strip of filter paper soaked with the suspected synergist is placed onto the plate (Fig. 2–12). If the agent is not synergistic a straight line of inhibition will result (A). If the agent is synergistic the zone of inhibition will extend along the strip (B). Quantitative evaluation of synergy is, however, not possible with this technique as the gradient of the potentiating agent is logarithmic and this overlaps the linear antibiotic gradient. Reversal of the position of the synergistic agent and the antibiotic may increase the detection of potentiation by this assay.

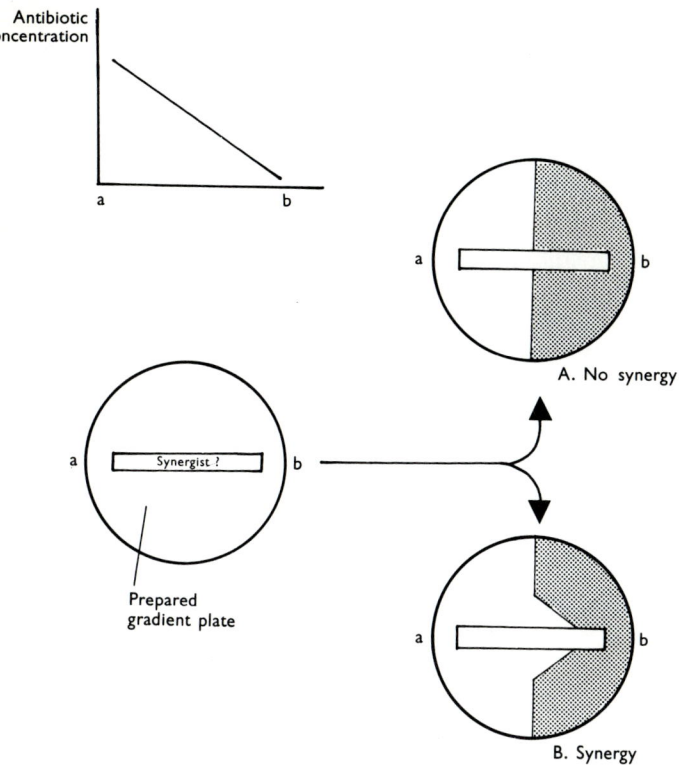

Fig. 2–12 The paper strip gradient test.

3 Inhibitors of Cell Wall Synthesis

3.1 The bacterial cell wall

The concentration of ions and metabolites inside bacterial cells is much greater than in the surrounding medium. The high concentrations are required for the cells to function normally: to generate energy, to synthesize macromolecules, and to grow and divide. This high intracellular concentration causes a high internal osmotic pressure. The delicate cell membrane that surrounds and contains the cytoplasm does not have sufficient strength to withstand such pressures and must be protected by the overlying cell wall.

The cell wall itself is a rigid structure composed of several macromolecules. The most important of these is peptidoglycan, a complex cross-linked network composed of sugar chains (glycans) made up from alternating N-acetylglucosamine and N-acetylmuramic acid residues. The N-acetylmuramic acid residues are substituted with a short chain of four amino acids (a tetrapeptide chain) and the glycan chains are joined together by bridges formed between these peptide chains (Fig. 3–1).

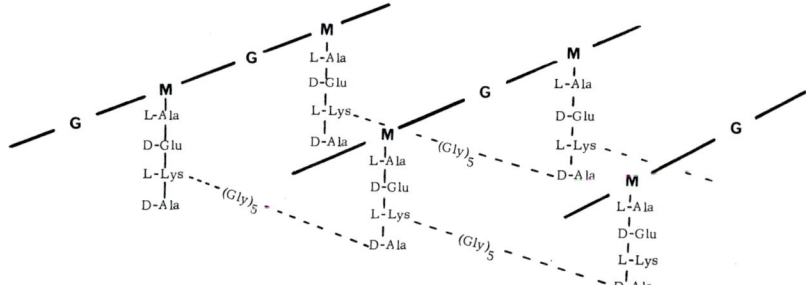

Fig. 3–1 Diagram showing the structure of the peptidoglycan in the cell wall of the bacterium, *Staphylococcus aureus*. Glycan chains are composed of N-acetylglucosamine (G) and N-acetylmuramic acid (M). Tetrapeptide chains attached to each M are cross-linked by peptide bridges containing 5 glycine residues. The amino acids are L-alanine (Ala), D-glutamine (Glu), L-lysine (Lys), D-alanine (Ala), and glycine (Gly).

The length of the glycan chains varies from organism to organism; in staphylococci the chains may contain only 12 units whereas in bacilli they may be up to 200 units long. Similarly the nature of the amino acids in the tetrapeptide chains and the way in which the cross-links are formed between the chains differ. However, all peptidoglycans have a similar gross macromolecular structure providing a cross-linked network of high tensile strength surrounding the delicate cell membrane and giving it mechanical support. The peptidoglycan acts rather like wire-netting, restraining the cell membrane which would otherwise burst under the influence of the high intracellular osmotic pressure.

3.2 Antibiotics affecting the cell wall

As peptidoglycan is unique to bacteria it is a sensitive site at which they can be selectively attacked. Agents which interfere with the biosynthesis of peptidoglycan and damage its cross-linked macromolecular structure can arrest growth and kill the organism. The more important antibiotics which inhibit peptidoglycan synthesis are listed in Table 1.

Table 1 Antibiotics affecting the cell wall

Antibiotic	Source	Major therapeutic applications
Penicillin	*Penicillium chrysogenum*	Gram-positive coccal infections, syphilis, gonorrhoea; meningococcal meningitis
Cephalosporin	*Cephalosporium spp.*	Used in place of penicillin for patients who are allergic to penicillin
Cycloserine	*Streptomyces orchidaceus*	Tuberculosis caused by drug-resistant bacilli
Bacitracin	*Bacillus licheniformis*	Sterilization of gut before surgery, topical application
Vancomycin	*Streptomyces orientalis*	Severe staphylococcal infections resistant to other drugs
Ristocetin	*Nocardia lurida*	Severe staphylococcal infections

Biosynthesis of peptidoglycan consists of three stages each of which occurs at a different site in the cell. The first stage, the synthesis of two uridine nucleotide precursors, UDP-acetylmuramyl-pentapeptide and UDP-acetylglucosamine takes place in the cytoplasm. The nucleotide (UDP, uridine diphosphate) itself does not form part of the wall, it is a means of bringing together the two basic building blocks of the wall—acetylmuramic acid and acetylglucosamine. As an analogy we might suppose that UDP is the wheelbarrow used to gather and assemble the bricks to build the wall. In the second stage the precursors are joined together to form a disaccharide pentapeptide, which is transported across the cell membrane to the wall assembly site. Because the lipid-rich cell membrane poses a barrier to the passage of the wall subunit, it is transferred to a lipid-soluble carrier molecule situated within the membrane. The disaccharide-pentapeptide moiety is transported across the cell membrane by the carrier lipid and delivered to the growing peptidoglycan on the outside of the membrane. Here the final stage takes place and consists of the cross-linking of the peptidoglycan strands. It is this cross-linking that gives the completed peptidoglycan its tensile strength and enables it to support the cell membrane.

The three stages in the synthesis of peptidoglycan are shown in Fig. 3–2

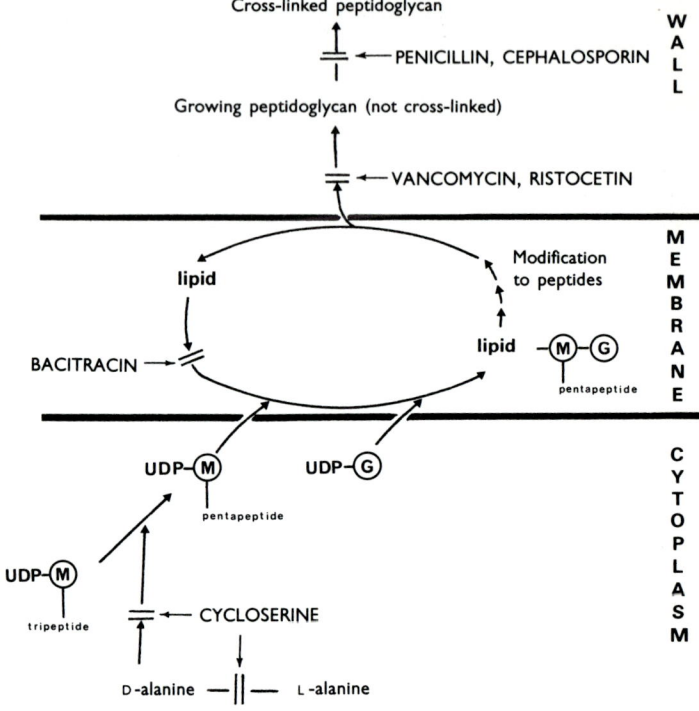

Fig. 3–2 Biosynthesis of peptidoglycan and some of the antibiotics which inhibit it. **M** represents N-acetylmuramic acid, **G** is N-acetylglucosamine and **lipid** is the lipid carrier molecule.

together with what are thought to be the specific sites of action of the antibiotics that affect cell wall synthesis.

3.3 Penicillins

The name 'penicillin' denotes a whole group of substances produced by penicillia and related moulds. Naturally occurring penicillins have a common ring structure, to which different acyl side-chains are attached as shown in Fig. 3–3.

The *Penicillium* strains first used in penicillin manufacture produced mainly penicillin F. It was found that addition of corn steep liquor (a nutritionally-rich extract from maize) to the growth medium not only improved the yield but also changed the nature of the major product to penicillin G. Corn steep liquor contains substances that can be used by the mould as precursors of phenylacetic acid, the acyl substituent of penicillin G. This experience showed that the nature of the penicillin produced can be markedly influenced by the addition to the medium of a

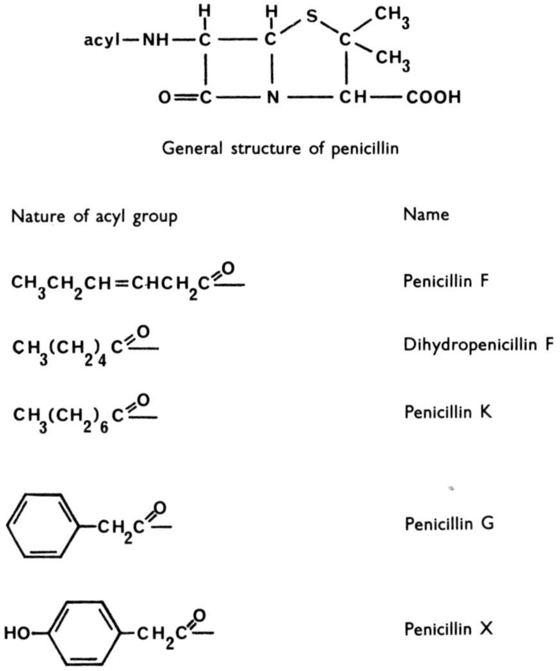

$$\text{acyl}-NH-\underset{|}{\overset{H}{C}}-\underset{|}{\overset{H}{C}}\overset{S}{\diagdown}\underset{CH_3}{\overset{CH_3}{C}}$$
$$O=C-N-CH-COOH$$

General structure of penicillin

Nature of acyl group	Name
$CH_3CH_2CH=CHCH_2C{\overset{O}{\diagup}}$	Penicillin F
$CH_3(CH_2)_4C{\overset{O}{\diagup}}$	Dihydropenicillin F
$CH_3(CH_2)_6C{\overset{O}{\diagup}}$	Penicillin K
⌬—$CH_2C{\overset{O}{\diagup}}$	Penicillin G
HO—⌬—$CH_2C{\overset{O}{\diagup}}$	Penicillin X

Fig. 3–3 The structure of some naturally-occurring penicillins.

biosynthetic precursor and, because penicillin G is more useful than penicillin F, the addition of phenylacetic acid to growing cultures became a routine part of the production process. The moulds are capable of incorporating many other compounds into the acyl side chain of the molecule provided the appropriate precursor is added to the medium. In this way, a very large number of new penicillins not normally produced can be synthesized biologically.

When the acyl group is removed from a penicillin by chemical or enzymatic hydrolysis, the product, 6-aminopenicillanic acid, shows no antibacterial activity. It can, however, be reacylated efficiently by chemical means which makes possible the production of a large variety of semi-synthetic penicillins.

When susceptible bacteria are grown in the presence of lethal concentrations of penicillin they lyse; if subinhibitory concentrations are used, large, swollen filamentous forms are produced. Such effects are observed only with growing bacteria. When rod-shaped organisms are grown in the presence of penicillin in a protective hypertonic medium containing magnesium ions and a stabilizing concentration of sucrose, large osmotically fragile, spherical forms called spheroplasts are produced. These bodies remain viable and will revert to rod-shaped organisms if penicillin is removed (Fig. 3–4).

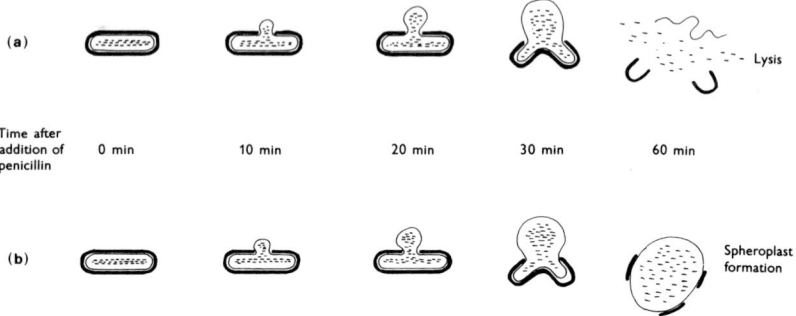

Fig. 3–4 Effect of penicillin upon growing cells of *Escherichia coli*: (a) in normal medium the cells lyse, (b) in a protective hypertonic medium spheroplasts are produced.

Penicillin specifically inhibits the last step in peptidoglycan synthesis, the cross-linking of the linear peptidoglycan strands. When bacteria are grown in the presence of penicillin, uncross-linked uridine nucleotide intermediates of peptidoglycan synthesis accumulate. These are known as 'Park nucleotides' after the discoverer J. T. Park; detection and analysis of Park nucleotides provides a useful approach in studying the mechanism of action of penicillin and other related antibiotics which act upon cell wall synthesis.

Penicillin is the most widely used antibiotic for general therapy, and penicillin G is the most useful of natural penicillins. At one time bacteria were classified as sensitive or resistant to penicillin, but they exhibit degrees of sensitivity over a wide range. Table 2 shows the concentrations of penicillin G necessary to inhibit the growth of some of the important Gram-positive and Gram-negative pathogenic organisms encountered in clinical practice.

Penicillin G is unstable in acid and the action of gastric acid accounts for the loss of most of a dose if it is swallowed. It can also be destroyed by

Table 2 Spectrum of activity of penicillin G

	Disease	Minimum inhibitory concentration ($\mu g/ml$)
Gram-positive organisms		
Clostridium tetani	tetanus	0·007–0·3
Staphylococcus aureus	boils, styes	0·012*
Corynebacterium diphtheriae	diphtheria	0·02–0·6
Gram-negative organisms		
Neisseria gonorrhoeae	gonorrhoea	0·003
Escherichia coli	gastro-enteritis	20
Salmonella spp.	enteric fever, typhoid	2–5

* Some strains are more resistant.

an enzyme, penicillinase, which is produced by various bacteria including staphylococci; the resistance of staphylococci to penicillin in clinical practice is due largely to penicillinase. For these reasons other penicillins are often preferred to penicillin G.

3.4 Cephalosporins

Cephalosporins are produced by a species of *Cephalosporium* isolated from the sea near a sewage outfall off Sardinia. Crude products from the organism were initially used with some success to treat typhoid fever and brucellosis. Close examination of the culture showed that it produces three distinct antibiotics designated cephalosporin N, P, and C. One of these, cephalosporin C, resembles penicillin in possessing a fused β-lactam ring (Fig. 3–5). Although it has only moderate antibacterial activity it possesses a high degree of resistance towards penicillinase. Cephalosporin C has been chemically manipulated in the same way as penicillin to give 7-amino cephalosporanic acid from which active derivatives such as cephalothin and cephaloridine have been prepared. These compounds are active against Gram-positive and Gram-negative bacteria and are resistant to staphylococcal penicillinase; like penicillin

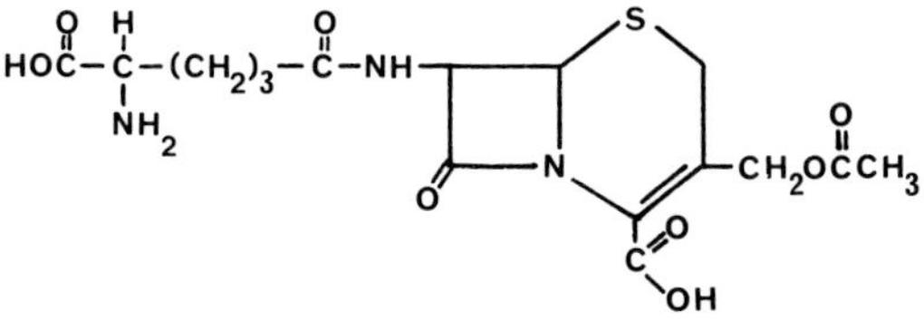

Fig. 3–5 The structure of cephalosporin C.

they are, however, susceptible to β-lactamases produced by certain Gram-negative bacteria. It is thought that cephalosporins have a similar mechanism of action to that of penicillin, the β-lactam ring is common to both molecules and may play a vital role in the antibacterial action.

3.5 Cycloserine

D-cycloserine is produced by *Streptomyces orchidaceus* and other organisms and has also been chemically synthesized. Its structural relationship to D-alanine is the basis of its antibacterial activity (Fig. 3–6). When *Escherichia coli* is grown in a hypertonic medium in the presence of

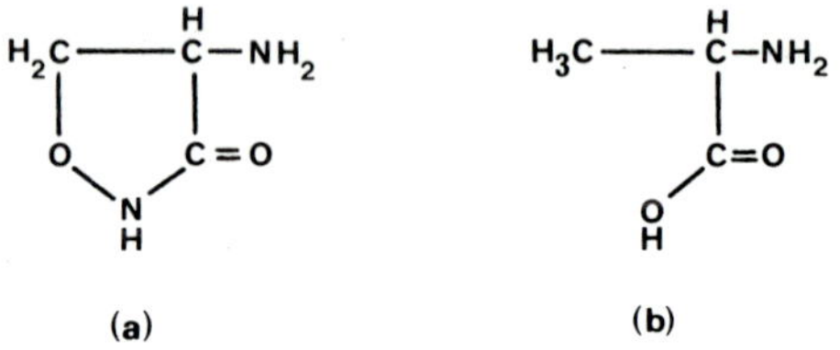

Fig. 3–6 The structural similarity between D-cycloserine (a) and D-alanine (b).

cycloserine, spheroplasts are formed. The inclusion of D-alanine in the medium inhibits spheroplast formation and prevents the inhibitory effect of cycloserine on growing cells. The accumulation of a uridine muramic acid peptide which occurs when organisms are grown in minimal inhibitory concentrations of cycloserine is also prevented by the addition of D-alanine. It is believed that cycloserine inhibits two of the initial steps in the synthesis of the peptidoglycan precursors involved in the incorporation of alanine into the pentapeptide chain. A wide range of Gram-positive and Gram-negative bacteria are sensitive to cycloserine and it has a low toxicity for man. It is active against *Mycobacterium tuberculosis* and has been successfully used in the treatment of tuberculosis.

3.6 Bacitracin

The name bacitracin is used for a group of closely related polypeptide antibiotics (designated bacitracin A, B and C, Fig. 3–7) produced by *Bacillus licheniformis*. Their mode of action resembles that of penicillin; they cause spheroplast formation and the accumulation of nucleotide precursors of peptidoglycan. The specific site of action is concerned with the lipid-soluble carrier molecule which transports the disaccharide-pentapeptide precursor across the cell membrane for assembly outside the membrane (Fig. 3–2). In the last step of this stage of peptidoglycan

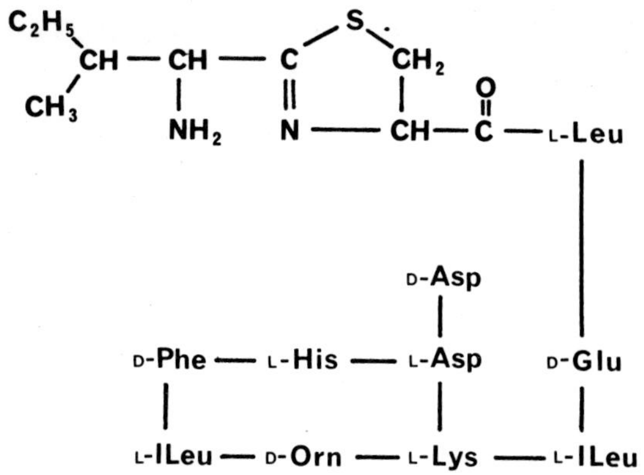

§ 3.7

$$
\begin{array}{c}
\text{C}_2\text{H}_5 \\
\text{CH} - \text{CH} - \text{C} \diagup^{\text{S}} \diagdown \text{CH}_2 \\
\text{CH}_3 \diagup \quad | \quad \quad \| \quad \quad | \quad \quad \text{O} \\
\quad \quad \text{NH}_2 \quad \text{N} - \text{CH} - \overset{\|}{\text{C}} - \text{L-Leu}
\end{array}
$$

```
                        D-Asp                |
                          |                  |
       D-Phe — L-His — L-Asp         D-Glu
         |              |              |
       L-ILeu — D-Orn — L-Lys — L-ILeu
```

Fig. 3–7 The structure of bacitracin A. Constituent amino acids are leucine (leu), phenylalanine (Phe), asparagine (Asp), histidine (His), ornithine (Orn), lysine (Lys) and glutamine (Glu).

biosynthesis the pyrophosphate form of the lipid-soluble carrier is dephosphorylated to yield inorganic phosphate and the carrier lipid. It is this last step that is specifically inhibited by bacitracin and the carrier lipid is thus prevented from re-entering the reaction cycle of peptidoglycan synthesis.

Bacitracin is highly active against many Gram-positive bacteria and the pathogenic *Neisseriae*, but as it is not absorbed from the stomach it cannot be given orally and it causes damage to the kidneys on injection. Consequently bacitracin is not used in clinical practice but it is an interesting and useful biochemical tool.

3.7 Vancomycin and ristocetin

Vancomycin is obtained from *Streptomyces orientalis*; it is a complex substance containing sugars and amino acids. It is a narrow-spectrum antibiotic active against Gram-positive bacteria and spirochetes. Vancomycin prevents the incorporation of amino acids into peptidoglycan when the disaccharide-pentapeptide is modified during the second stage of peptidoglycan synthesis. The transfer of the linear peptidoglycan from the lipid carrier to the growing peptidoglycan outside the membrane is also blocked.

Ristocetin is a mixture of two closely related substances produced by *Nocardia lurida* it is composed primarily of sugars and amino acids. It has pronounced bactericidal activity for Gram-positive bacteria and mycobacteria, and has been successfully used in the therapy of severe staphylococcal infections. The mechanism of action of ristocetin appears to be similar to that of vancomycin in preventing transfer of the disaccharide-pentapeptide units from the lipid carrier.

4 Membrane-active Antimicrobial Agents

4.1 Structure and function of cell membranes

The cell membrane forms an expandable barrier to the microbial protoplast, acts as an organelle controlling the entry and exit of solute molecules and provides a site for synthesis at which cell wall and extramural layers are assembled. These many functions make the membrane vulnerable to a variety of agents. Disturbance of membrane function is often lethal. Several groups of antimicrobial agents produce their effect by acting upon the cell membrane, including antiseptic compounds (e.g. cetrimide), the peptide antibiotics (e.g. polymyxin) and the polyene antifungal antibiotics.

Cell membranes are asymmetric lipid bilayers in which cytoplasmically synthesized proteins are dissolved. The constitutent lipid (predominantly phospholipid) molecules consist of two parts, a polar (or charged) hydrophilic (water-loving) head group and a hydrophobic (water-hating) fatty acid tail. This combination of hydrophobic and hydrophilic regions gives phospholipids novel properties when dissolved in aqueous systems, the hydrophobic region preventing the formation of true solutions. Instead, phospholipids form multimolecular aggregates called micelles in which the polar hydrophilic head groups interface with the aqueous environment and the fatty acid tails combine to form a localized hydrophobic region. The phospholipids of cell membranes are arranged in a bilayer, with the hydrophilic head groups pointing outwards and the lipid tails creating a hydrophobic central region.

Many models have been proposed to explain the distribution of protein components in the lipid bilayer. Figure 4–1 gives what many, but not all, believe to be the generalized structure of bacterial membranes. Certain proteins are located in the outer half of the membrane, some in the inner half, while others completely traverse the membrane. Membranes must not be considered as fixed structures, indeed they are very fluid. Within each half of the bilayer both lipid and protein components move, sometimes extremely rapidly. However, this movement is restricted to within each half of the bilayer. Movement of lipid and protein from one half of the bilayer to the other occurs only rarely.

4.2 Membrane-active antiseptics

Membrane-active antiseptics can be divided into two groups. The first group, the phenolic compounds (aromatic molecules containing one or

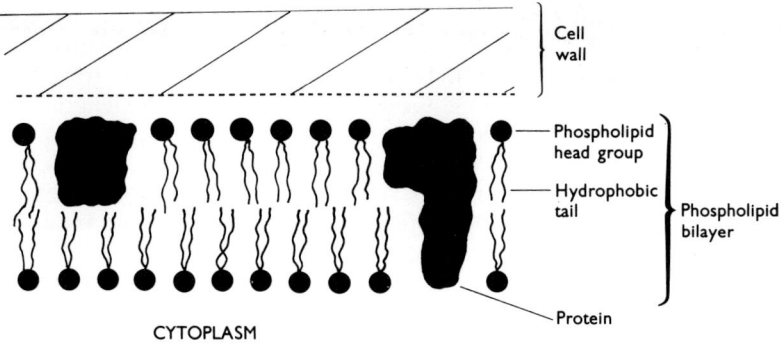

Fig. 4–1 Diagram showing the generalized morphology of the microbial cell surface.

more hydroxyl groups directly linked to a benzene ring), includes the well known chlorinated cresols and xylenols, e.g. Dettol; chlorinated phenols, e.g. TCP; and chlorinated biphenols, e.g. hexachlorophene (Fig. 4–2). The second group, the cationic antiseptics, includes a range of molecules of

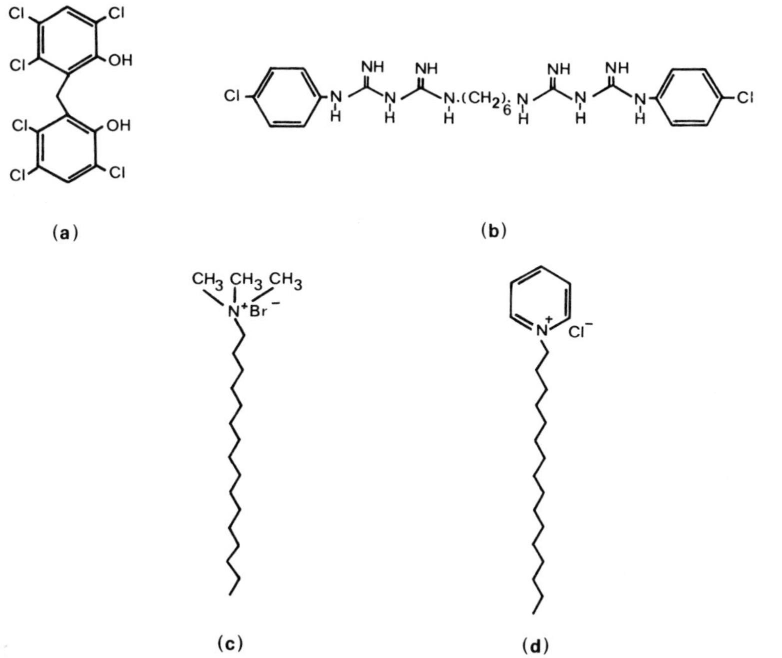

Fig. 4–2 The structure of some common antiseptics: phenolic antiseptic (a) hexachlorophene; cationic antiseptics (b) chlorhexidine, (c) CTAB, (d) CPC.

different chemical types but which all contain a strong basic group attached to a long alkyl chain. The most important members of this group are cetyltrimethylammonium bromide (CTAB, sold as Cetrimide or Savlon) and cetylpyridinium chloride (CPC) (Fig. 4–2). The length of the constituent alkyl chain is critical to the antimicrobial properties of these compounds; the synthesis of molecules, with shorter or longer chains changes the antimicrobial activity. Another important cationic antiseptic is chlorohexidine (Hibitane) which has two strongly basic biguanide groups (Fig. 4–2).

The primary lesion induced in the microbial cell by the antiseptic is located in the cell membrane. When bacteria are treated with lethal antiseptic concentrations, the antimicrobial agent is found to be exclusively bound to components of the cell membrane, probably proteins. If the bacterial cell wall is removed to make protoplasts or membrane fragments, these become sensitive to very low antiseptic concentrations, being readily damaged or lysed. When the ability of the cell membrane to act as a semi-permeable barrier between the cell and the environment is impaired by the binding of antiseptic molecules, low molecular weight cytoplasmic components begin to appear in the surrounding medium. Microbial cells accumulate potassium ions against a concentration gradient, i.e. they actively take up this ion from the environment to achieve intracellular concentrations many times greater than that in the environment. The speed at which cytoplasmic components leak from cells after the membrane has been damaged is inversely related to their molecular size. Because of the smallness and high internal concentration of potassium ions, the first observable sign of membrane damage is the leakage of these ions from the cell. This is followed by the appearance in the medium of material recognized by its ability to absorb light of wavelength 260 nm in a spectrophotometer, the so-called 260 nm-absorbing material, consisting of nucleotides and certain amino acids. Phosphate ions and sugars also leak from the cell. Although potassium is necessary for many cellular reactions and its loss may inhibit metabolism, the leakage of cellular material is only indicative of gross morphological and biochemical changes induced in the membrane by the binding of the antiseptic. The modification of the membrane by the drug may release cellular enzymes capable of degradation of protein and nucleic acids eventually leading to lysis. Treatment with very high antiseptic concentrations produces extensive non-specific damage to cell morphology and biochemistry resulting from penetration of the antiseptic into the cell interior.

The use of antiseptics in medicine is not restricted to the cleansing of minor wounds; these compounds, particularly the cationic antiseptics, are used extensively in the therapy of skin and mucous membrane infections and as the antimicrobial principle in throat lozenges, mouthwashes and skin creams.

4.3 Cyclic polypeptide antibiotics

The cyclic polypeptide antibiotics can be divided into two groups on the basis of the number of amino acids constituting the polypeptide ring. Members of the first group which includes the tyrocidins and gramicidin S are cyclic decapeptides (deca = 10) containing one, or occasionally two, free amino groups (Fig. 4–3). The second group, the polymyxins, consists of a series of chemically related cyclic antibiotics produced by *Bacillus* species. The polymyxins have a smaller polypeptide ring (7-membered) attached to a short polypeptide chain terminated by a branched fatty acid (Fig. 4–3). Treatment of bacteria with any of the cyclic polypeptides produces effects very similar to those elicited by the membrane-active antiseptics described above.

Low tyrocidin concentrations are bactericidal to Gram-positive organisms by combining to the cell membrane, disturbing membrane function, allowing leakage of cytoplasmic components and uncoupling oxidative phosphorylation (i.e. preventing the generation of ATP during sugar oxidation). The chemical structure and antimicrobial action of

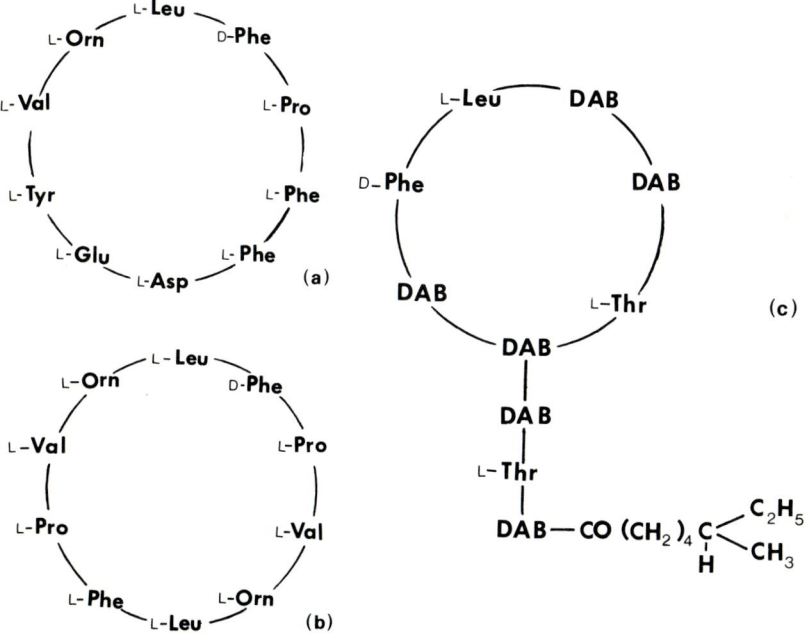

Fig. 4–3 Cyclic polypeptide antibiotics (a) tyrocidin A, (b) gramicidin S, and (c) polymyxin B1. Constituent amino acids are leucine (Leu), phenylalanine (Phe), proline (Pro), asparagine (Asp), glutamine (Glu), tyrosine (Tyr), valine (Val), ornithine (Orn), and 2,4-diaminobutyric acid (DAB).

gramicidin S is very similar to tyrocidin. Because of high toxicity to mammals tyrocidin and gramicidin S have found little use in medicine, although tyrocidin is often incorporated in throat lozenges.

Members of the polymyxin series bind strongly and exclusively to the phospholipid component of bacterial membranes destroying both membrane integrity and function inevitably leading to death. Polymyxin can be used systemically and is particularly useful as it is one of the few antibiotics effective against heavy Gram-negative (notably *Pseudomonas*) infections, although there is a risk of its causing kidney damage.

4.4 Ionophore antibiotics

This group contains both cyclic and linear polypeptide antibiotics, which differ from the previous group in that they have the ability to facilitate the passage of inorganic ions across membranes (ionophore=ion bearer). These compounds are not used clinically, as they are equally toxic to mammalian and bacterial cells. However, they exhibit several novel properties which deserve further description. Valinomycin, the ionophore which has received the most attention, is a cyclic depsipeptide, in which the twelve components of the ring are arranged in alternating pairs of D and L configurations (Fig. 4–4). This antibiotic shows

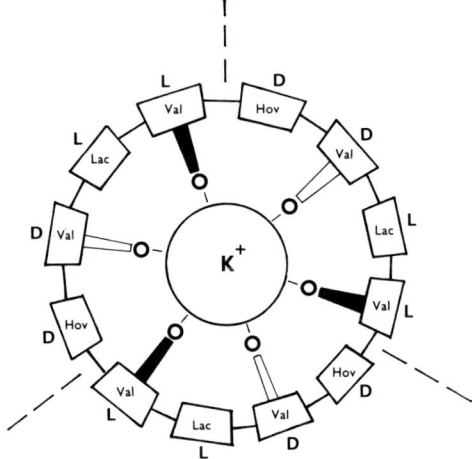

Fig. 4–4 Structure of the valinomycin-potassium complex. In the valinomycin ring the three residues are valine (Val), 2-hydroxy-isovaleric acid (Hov), and lactic acid (Lac). The dotted lines separate the repeating units and the asymmetric centres are labelled D or L. The central potassium cation (K^+) is co-ordinated by 6 oxygen atoms derived from carbonyl groups of valine residues. The three dimensional nature of the valinomycin ring is represented by imagining the solid L-valine ligands as coming towards the reader and unshaded D-value ligands passing into the page.

high affinity for potassium ions. X-ray diffraction and other studies indicate that in the potassium-valinomycin complex a single potassium ion is completely surrounded by the antibiotic molecule and is tightly held by hydrogen bonds. The ability to form a complex depends entirely upon the alternation of D and L pairs; any change in the conformation of the molecule destroys the ability to form a complex. The dimensions of a potassium ion exactly fit the hole in the centre of the valinomycin ring. Other cations, e.g. sodium can also complex with valinomycin but their smaller size makes such complexes less stable (by a factor of 10^3). Valinomycin facilitates the transport of potassium across cell membranes. This is thought to result from the movement of the lipophilic potassium-valinomycin complex from the cytoplasm/membrane interface, through the membrane to the outside, where the potassium ion is discharged to the environment. The valinomycin molecule would then be free to return to the inner side of the membrane to pick up another ion. The flow rate depends upon the internal potassium concentration; while this remains greater than that of the environment, valinomycin will drain the cell of its potassium. It is believed that the loss of cellular potassium prevents energy production and protein synthesis leading to death.

Several other ionophores have been discovered. Some of their properties are given in Table 3.

Table 3 Properties of some ionophorous antibiotics

Antibiotic	Chemical nature	Ion specificity	Ratio, antibiotic molecules : ion
Valinomycin	cyclic depsipeptide	$K^+ > Na^+ \gg Ca^{++}, Mg^{++}$	1 : 1
Monactin	cyclic macrotetralide	$K^+ > Na^+ \gg Ca^{++}, Mg^{++}$	1 : 1
Monensin	linear macrotetralide	$Na^+ > K^+ \gg Ca^{++}, Mg^{++}$	1 : 1
Nigerisin	linear macrotetralide	$K^+ > Na^+ \gg Ca^{++}, Mg^{++}$	1 : 1
A23187	linear macrotetralide	$Ca^{++} = Mg^{++} \gg K^+, Na^+$	1 : 2

4.5 Gramicidin A

This antibiotic should not be confused with the unrelated antibiotic gramicidin S. It is a linear polypeptide and exhibits antimicrobial properties which are very similar to those of valinomycin. It shows a high affinity for potassium ions and facilitates their passage across lipid membranes. However, it has been demonstrated that gramicidin A does not act as an ionophore. Valinomycin only acts as a potassium carrier when the membrane is fluid. If the membrane is cooled to a point when the lipid components cease to be mobile, valinomycin is unable to move across the membrane to transport potassium. In contrast, gramicidin A continues to facilitate potassium movement at low temperatures. It is

34 POLYENE ANTIFUNGAL ANTIBIOTICS § 4.6

believed that gramicidin A forms a helical cylinder in lipid membranes with lipophilic groups on the outside of the cylinder and hydrophilic groups lining a central hole (Fig. 4–5). This creates a hydrophilic pore across the cell membrane able to conduct potassium ions from the cell interior to the environment. It has been calculated that a single such channel could convey 3×10^7 potassium ions/second.

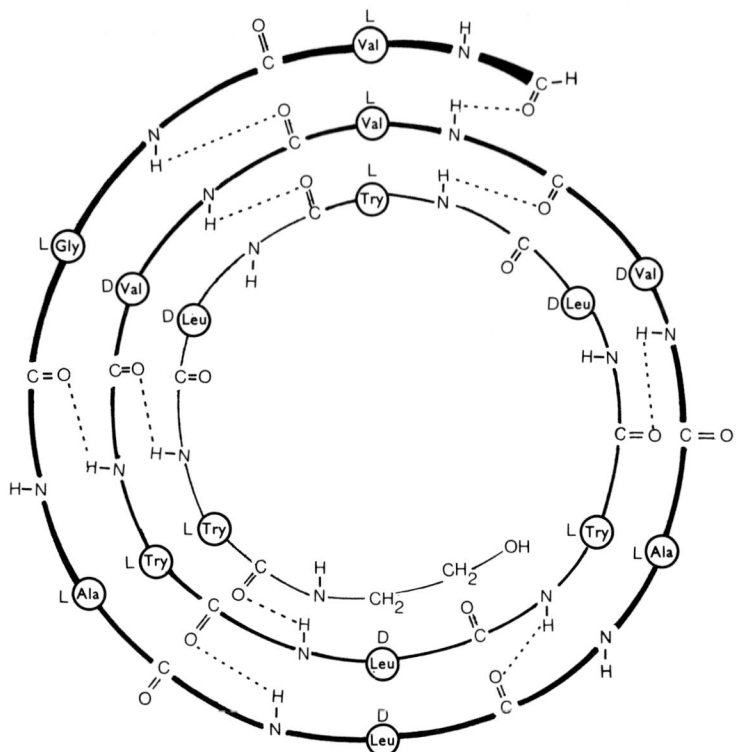

Fig. 4–5 A possible helical structure for gramicidin A with 6.3 residues for each turn of the helix. Readers should imagine that they are looking down a 'tunnel' with parallel sides; those bonds which are drawn inwards being directed down the helix and those drawn outwards being directed upwards. The asymmetric centres of the constituent amino acids alanine (Ala), glycine (Gly), leucine (Leu), tryptophane (Try), and valine (Val) are marked either D- or L-.

4.6 Polyene antifungal antibiotics

The polyene antibiotic group consists of a series of macrocyclic ring compounds produced by *Streptomyces* sp., which although potent antifungal agents show no antibacterial activity. Polyene antibiotics are used extensively in the control of pathogenic and opportunistic fungal infections. (Opportunistic infections are caused by organisms which normally exist in a state of balance with the host causing little or no damage, but which may become pathogenic if this balance is disturbed.) In humans mycotic (=fungal) infections are generally confined to the skin, e.g. athlete's foot, ringworm; to moist mucous membranes, e.g. oral thrush, candidial vaginitis; and intestinal infections. The polyene antibiotics are relatively toxic when injected into the bloodstream, but as they are poorly absorbed by skin and mucosal surfaces, polyenes can be used topically. Although systemic mycoses (infections where the fungus invades the bloodstream and viscera) occur more rarely than superficial mycoses they are nearly always fatal. In such cases the risks of intravenous polyene therapy, mainly kidney damage and haemolysis, are outweighed by the severity of the infection. Of the fifty or so polyene antibiotics that have been described, nystatin is generally used for topical application and amphotericin B for systemic therapy. Other polyenes, notably candicidin and pimaricin have also been used in antifungal therapy. Because of their toxicity to mammals the remaining polyenes have little clinical use.

Polyene antibiotics are characterized by the possession of a large macrolide ring (Fig. 4–6) consisting of two very distinct parts. The first is the series of conjugated double bonds which gives the group its name (poly=many, -ene=double bond). Different polyenes can have from four to seven conjugated bonds, which are strongly hydrophobic and impart rigidity to the molecule. The carbon chain on the opposite side of the ring contains a large number of hydroxyl groups creating a flexible hydrophilic region. The polyene molecule also contains an amino sugar, mycosamine.

It has been demonstrated that polyene antibiotics will bind only to those membranes which contain sterols. Sterols exhibit a restricted distribution in nature being necessary for the structure and function of the membranes of eukaryotes (e.g. fungi, green plants and animals) but absent from the membranes of prokaryotic organisms (e.g. bacteria and blue-green algae). When eukaryote cells are treated with polyene

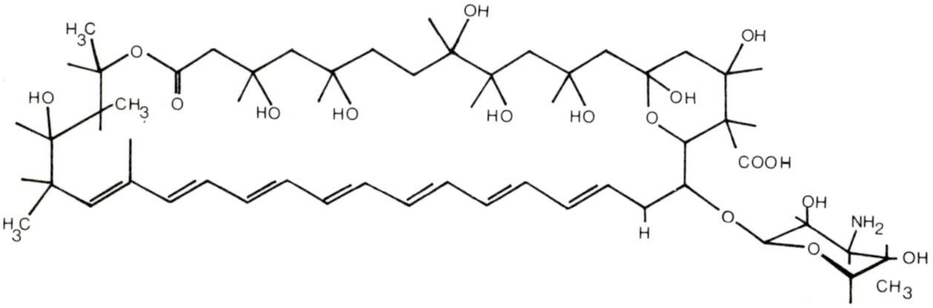

Fig. 4–6 The structure of amphotericin B, a typical polyene antifungal antibiotic.

antibiotics the cell membrane rapidly loses its ability to act as a selective permeability barrier between the cell contents and the environment leading to the leakage of potassium ions and small molecules. It is believed that when the antibiotic molecule reaches the cell membrane, the strong affinity of the hydrophobic region of the polyene for the membrane sterol drags the antibiotic molecule into the membrane, so that the hydrophobic region lies alongside the sterol ring. This not only creates instability in the membrane by breaking the phospholipid-sterol interactions responsible for stabilizing the membrane, but also introduces the hydrophilic polyol surface of the antibiotic into the membrane interior (Fig. 4–7). The intrusion of a hydrophilic chain into the membrane may be sufficient to account for the observed polyene-induced permeability changes. It has been suggested that several sterol-polyene complexes come together to form aggregates with their hydrophilic regions arranged to create a central hydrophilic pore in the membrane (Fig. 4–7).

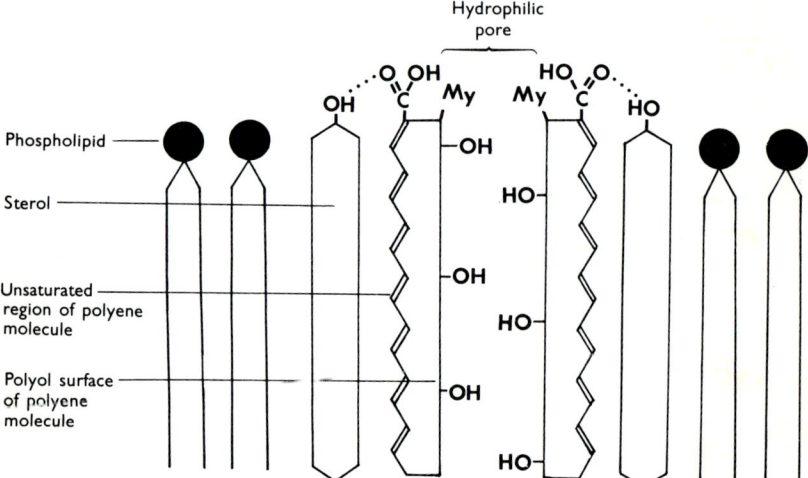

Fig. 4–7 Postulated structure for the polyene-sterol pore formed in eukaryote membranes (My=mycosamine).

5 Inhibitors of Nucleic Acids

5.1 Nucleic acids

Genetic information in all living cells is carried by the chromosomes. The region of a chromosome that determines a particular character is called a gene. Chemically, genes are composed of deoxyribonucleic acid (DNA). The structure of all enzymes produced by cells and the rates at which they are made are directly controlled by the genes. The expression of the gene, that is the determination of protein structure and synthesis, involves several other forms of nucleic acid. These are ribonucleic acids (RNA) and the three forms that occur in the cell are designated as messenger RNA (mRNA), ribosomal RNA (rRNA), and transfer RNA (tRNA).

5.2 Structure of DNA and RNA

DNA and RNA consist of long chains of alternating sugar and phosphate residues with organic bases (purines or pyrimidines) attached to each sugar unit. DNA contains 2-deoxyribose as the sugar residue and the various forms of RNA contain ribose. The major purine and pyrimidine bases are adenine, guanine, and cytosine, which occur in both DNA and RNA, together with thymine, which occurs exclusively in DNA, and uracil, which occurs exclusively in RNA. The molecular structure of DNA was first described by J. D. Watson and F. H. C. Crick in 1953 (Fig. 5-1). They showed that DNA consists of two polynucleotide chains running in opposite directions and wrapped around each other to form a helix with the sugar-phosphate chain on the outside and the purine and pyrimidine bases on the inside. The chains are held together by hydrogen bonds formed between the bases in either chain. The hydrogen bonding is highly specific and base pairs cannot be formed in any combination other than between adenine and thymine, and between guanine and cytosine. Consequently the sequence of bases in one of the chains determines absolutely and unambiguously the sequence of bases that must occur in the opposite (complementary) chain. This factor is of fundamental importance to the function of DNA in living systems.

In comparison with DNA the three forms of RNA do not possess such highly-ordered structures. They exist generally as single polynucleotide chains which may be folded over in a number of places to form loops.

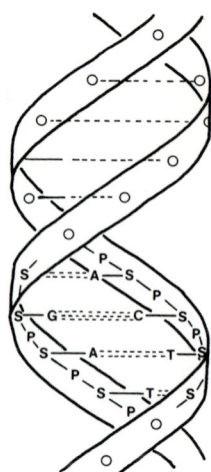

Fig. 5–1 The double helical structure of DNA. S=deoxyribose, P=phosphodiester, A=adenine, C=cytosine, G=guanine, T=thymine. Hydrogen bonding between bases is represented by dotted lines.

5.3 Biological roles of DNA and RNA

DNA is an informational molecule. The information it contains is genetic: the directions for the production of specific proteins and information for the propagation of the species in a relatively unvarying manner. DNA has two general functions: (1) it must provide directions for self-duplication during cell division so that the information it carries can be transmitted to daughter cells; and (2) it must express its encoded information for controlling the metabolic activity of the cell. These functions are mediated respectively by DNA biosynthesis (replication) and mRNA biosynthesis (transcription).

The various cellular forms of RNA (transfer, messenger, and ribosomal) all play important roles in the translation of the genetic information contained in the DNA. We shall deal with the mechanism of protein synthesis and its inhibition by antibiotics in the next chapter; in this chapter we are concerned with the events that take place around the DNA molecule, i.e. the processes of replication and transcription.

5.4 Semi-conservative replication of DNA

The replication of DNA molecules is thought to occur as shown in Fig. 5–2. The two polynucleotide strands of the DNA double helix must first separate over at least a short region of the molecule thereby exposing the base sequence on one of the chains and the complementary sequence on

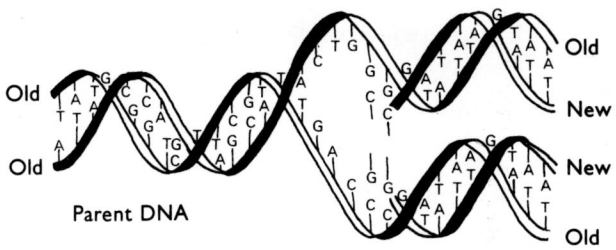

Fig. 5–2 Semi-conservative replication of DNA. The strands of the parent DNA partly unwind permitting new strands to be synthesized along each of the old strands.

the other chain. Free deoxyribonucleotides in the cell can then become attached by hydrogen bonds to the exposed bases on the separated strands. the nucleotides are then linked together by a polymerizing enzyme, DNA polymerase, and two new molecules of DNA are formed each containing one strand from the original DNA molecule and one newly-synthesized strand. For this reason the process is called semi-conservative replication and it ensures that the correct sequence of bases in the DNA, constituting the genetic code, is maintained.

5.5 Transcription and RNA biosynthesis

The first process along the path of information flow from DNA to protein is transcription of the base sequence on the DNA. This is achieved by synthesizing mRNA with a base sequence complementary to that of a section of the DNA molecule. The process is similar to that of DNA replication: a DNA-dependent RNA polymerase joins together ribonucleotides in the prescribed sequence using one of the DNA strands as a template.

5.6 Antibiotics which affect nucleic acid function

Although they exhibit antimicrobial activity many of the antibiotics we shall discuss are used in cancer chemotherapy rather than as antimicrobial agents. Present-day cancer chemotherapy relies very largely upon inhibition of nucleic acid synthesis as a means of discriminating selectively against the fast-growing tumour cells in the body, leaving the slower-growing healthy cells undamaged. The pressing need for effective treatment of cancer has resulted in the investigation of a wide variety of compounds which interfere with nucleic acid synthesis and function. Unfortunately many of these do not show sufficient selectivity against the tumour cells and are too toxic to be used effectively.

5.7 Agents which interfere with nucleotide biosynthesis

Azaserine and DON (6-diazo-5-oxo-L-norleucine) are related antibiotics produced by *Streptomycetes* which inhibit one particular step in the pathway of purine biosynthesis. Azaserine was the first antibiotic to be discovered as a direct result of a systematic search for tumour-inhibiting antibiotics. Both azaserine and DON are structural analogues of the amino acid glutamine (Fig. 5–3) and it is thought that their inhibitory

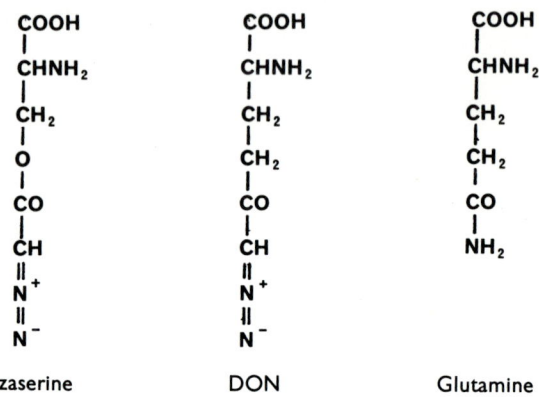

Fig. 5–3   The structural similarities between azaserine, DON and the amino acid glutamine.

action is due to the fact that they compete with glutamine for the binding site on an enzyme which is involved in the purine biosynthetic pathway. Once bound to the enzyme they inhibit it irreversibly. Azaserine appears to form a covalent bond with an -SH group on the enzyme.

5.8 Agents which interfere with the polymerization of nucleotides by impairing the template function of DNA

We have seen that DNA acts as a template molecule in the processes of replication and transcription. These functions can be impaired by certain antibiotics in two ways: the antibiotic molecules can bind to the DNA and form a complex directly inhibiting its function; alternatively the inhibition may be indirect, resulting from damage to the DNA such as strand breakage, removal of bases, or formation of cross-links between the strands. The most important mechanism is that of complex formation; complexes are usually formed by insertion ('intercalation') of drug molecules between the adjacent base pairs of the DNA double helix.

§ 5.8 INTERCALATING DYES

5.8.1 The intercalating dyes: acridines and ethidium

Although strictly speaking they are not antibiotics, the acridine dyes (Fig. 5–4) were among the earliest antibacterial agents known. The use of acridines as specific stains for nucleic acids in histology gives a good indication that they bind selectively and tightly to nucleic acids. The acridine proflavine was first used as a topical disinfectant on wounds during the First World War, but is much too toxic to be used internally as a systemic antibacterial agent. Ethidium bromide is a phenanthridine compound which binds tightly to DNA and is active against bacteria.

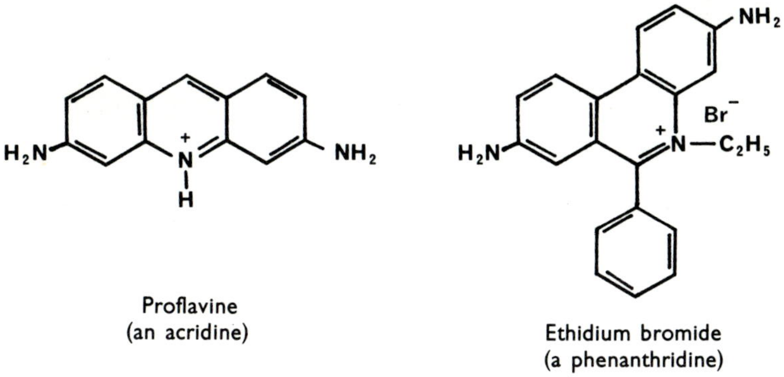

Proflavine
(an acridine)

Ethidium bromide
(a phenanthridine)

Fig. 5–4 The structures of two dyes that intercalate between the strands of DNA.

It is thought that the flat aromatic proflavine molecules become inserted (i.e. intercalated) between adjacent base-pairs in the double helix as shown in Fig. 5–5. The hydrogen bonding between base-pairs remains undisturbed although there is some distortion of the smooth helical coils of the sugar-phosphate chains. The intercalated drug molecules are probably held in place by electronic interaction with the bases of the DNA above and below them. A similar model has been described for the interaction between ethidium bromide and DNA in which hydrogen bonds can be formed between amino groups of the drug molecules and charged oxygens of the phosphate groups of both polynucleotide chains.

The intercalating compounds inhibit both DNA synthesis and DNA-dependent RNA synthesis in microbial cells. Since strand separation is an important event in the replication of DNA the increased stability of the DNA double helix resulting from intercalation of drug molecules could be an important factor in the inhibition of DNA replication. The intercalated drug molecules might also hinder the attachment of the enzyme RNA polymerase to the DNA template and thereby markedly reduce the rate of RNA synthesis.

5.8.2 Actinomycin D

Actinomycin D (Fig. 5–6) is perhaps the best known and most widely

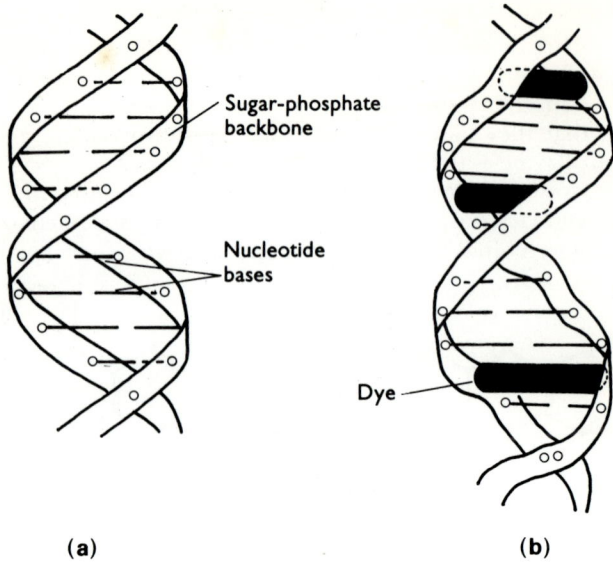

Fig. 5–5 Mechanism of antimicrobial activity of intercalating dyes: (a) represents the normal structure of DNA while (b) shows the distortion of the sugar phosphate backbone of the DNA helix created by the intercalated dye molecules.

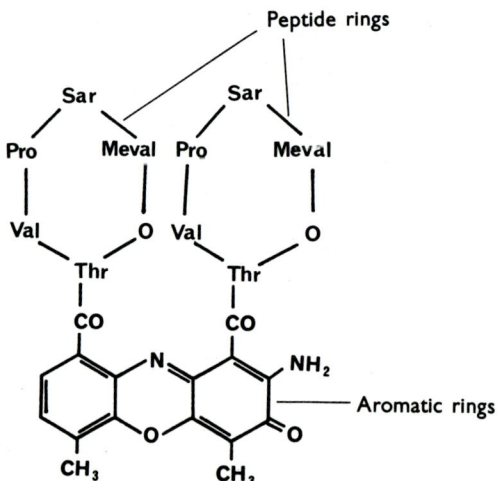

Fig. 5–6 The structure of actinomycin D. Constitutent amino acids are L-threonine (Thr), D-valine (Val), L-proline (Pro), sarcosine (Sar) and L-N-methylvaline (Meval).

studied of the antibiotics which bind to DNA. It is highly toxic and has no therapeutic value as an antimicrobial agent although it has found some limited clinical use as an anti-cancer agent.

A number of models have been proposed to explain the interaction of actinomycin D with DNA. One of these suggests that the actinomycin molecules insert themselves into the smaller groove of the DNA double helix (Fig. 5–7) where they are bound by hydrogen bonds.

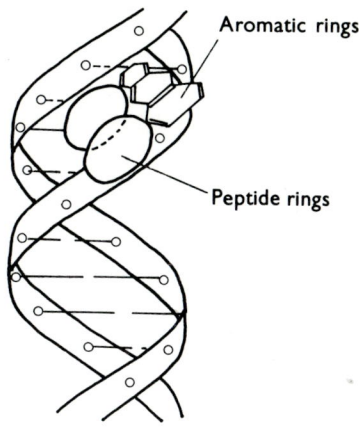

Fig. 5–7 Diagram showing how actinomycin D lies in the so-called minor groove of double-stranded DNA.

5.8.3 *Cross-linking of DNA: mitomycin C*

Mitomycin C is an antitumour antibiotic produced by various species of *Streptomyces* which is also active against bacteria. Inhibition of bacteria by mitomycin coincides with cessation of DNA synthesis and degradation of existing DNA while RNA and protein synthesis continue. Mitomycin exerts its effect by joining together the two polynucleotide strands of DNA forming a cross-link. The cross-link consists of a mitomycin molecule covalently bonded to two sites, one on each strand. The number of cross-links formed in a single DNA molecule is very low, probably as low as one cross-link per molecule but is sufficient to interfere with DNA function and inhibit DNA synthesis.

5.8.4 *Strand-breaking of DNA: streptonigrin*

Streptonigrin is an antitumour antibiotic which bears some structural similarity to mitomycin C but has the opposite affect upon DNA. Its lethal action is associated with degradation of DNA; after short exposure to streptonigrin bacterial DNA is found to contain numerous breaks.

5.9 Agents which interfere with enzymes involved in nucleic acid synthesis

Few of the antibiotics that act by binding to DNA are of much therapeutic value, as they are generally incapable of discriminating between the DNA of different types of cell, i.e. they show no selective toxicity. Since all cells contain DNA they are all potentially susceptible to inhibition by such antibiotics. Any useful selectivity that does exist is usually caused by other factors such as differing permeability of the cells towards the drug molecules. By way of contrast the antibiotics we shall discuss in the last section of this chapter do exhibit selective toxicity; they inhibit nucleic acid synthesis not by binding to DNA but by interfering specifically with enzymes involved in the polymerization reactions.

5.9.1 Inhibitors of RNA polymerase: rifamycins

The rifamycins are a group of antibiotics produced by *Streptomyces mediterranei*. They are active against Gram-positive bacteria and *Mycobacterium tuberculosis* but less active against Gram-negative bacteria. Rifampicin is a semisynthetic derivative of the naturally occurring rifamycin B which exhibits greatly increased antibacterial activity over natural rifamycins. Its bactericidal action correlates closely with the selective inhibition of RNA synthesis. It binds to, and inhibits the DNA-dependent RNA polymerase of sensitive bacteria. However, the corresponding enzymes from rifampicin-resistant strains of bacteria and from mammalian cells do not bind rifampicin and are not inhibited by it. The binding of the antibiotic to the enzyme is a key factor in the inhibition but the precise mechanism of inhibition is still uncertain.

6 Inhibitors of Ribosome Function

6.1 Protein synthesis: the function of the ribosome

Proteins are polymers made up from chains of amino acids joined together by peptide bonds. Twenty different amino acids can occur in proteins and it is the chemical nature and the order of the amino acids in the polymers that give them their specific properties and functions in the cell. They serve many different purposes in the cells, some are structural proteins which help to maintain the shape and integrity of cell components such as the cell membrane; others act as enzymes controlling the metabolic activity of the cells. During the normal growth and division of cells many different proteins are required and they must be synthesized in a controlled way: at the right time, in the correct quantity and have the chemical structure necessary for them to carry out their function. All the information for the controlled production of protein is stored in the genes, encoded in the sequence of bases in the DNA.

In order to produce a particular protein molecule the genetic code on that part of the DNA which carries the necessary information is first copied (transcribed) onto a newly-synthesized strand of messenger RNA. The information on the mRNA is then translated into the specified protein molecule at a site in the cell that is specially designed for the purpose—the ribosome.

Ribosomes are roughly spherical particles made up from two subunits each composed of protein and ribosomal RNA. They are found in large numbers in the cytoplasm of all cells: animal, plant and microbial. The sequence of events by which protein molecules are synthesized is shown in Fig. 6–1. The mRNA produced by transcription of part of the DNA (1) becomes attached to a ribosome (2). Molecules of transfer RNA supply the amino acids required for the particular protein (3). Each tRNA binds to the mRNA by hydrogen bonding between complementary bases in either molecule. While bound to the mRNA the amino acid carried by the tRNA is added to the growing peptide chain on the ribosome by formation of a peptide bond (4). When the correct number of amino acids has been assembled, the new polypeptide chain is released from the ribosome and assumes the configuration necessary for its particular function. The overall process is called translation since the information stored in the genes in code form is translated into protein. The sequence of amino acids in each new protein is controlled by the sequence of bases in the mRNA, which in turn is determined by the DNA. Each amino acid is supplied by a different tRNA molecule which possesses a sequence of three bases (the anticodon) enabling it to bind to a complementary

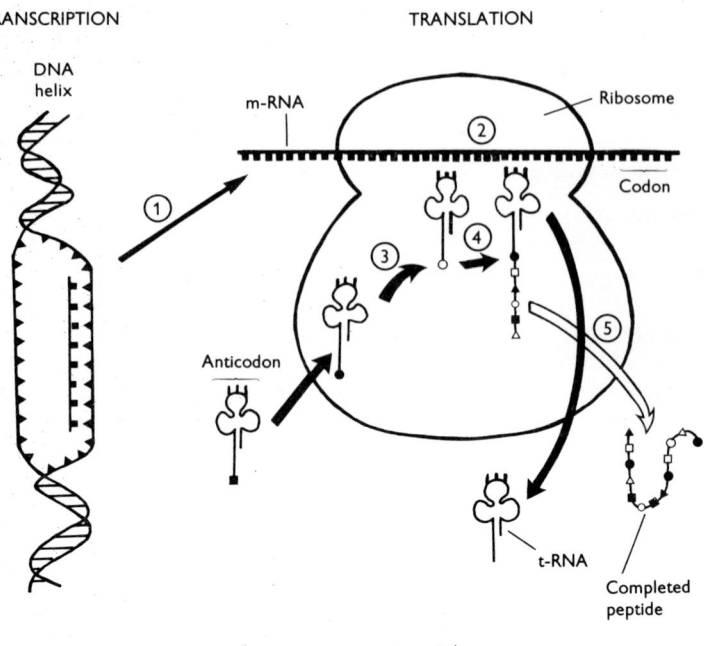

Fig. 6–1   Protein synthesis.

sequence of three bases (the codon) on the mRNA. As each amino acid is incorporated the mRNA strand moves along relative to the ribosome so that the next codon specifying the next amino acid can attract and bind the tRNA bearing that amino acid.

The mechanism of protein synthesis is essentially the same in all living cells, the only major difference is the size of the ribosomes and the subunits from which they are made. The ribosomes of mammalian and plant cells are slightly larger than those of bacteria. Some of the antibiotics which affect the function of the ribosome are only active against the smaller bacterial ribosomes, e.g. chloramphenicol and streptomycin. Others (e.g. tetracycline) inhibit the function of both kinds of ribosome when studied in cell-free systems but are far more effective inhibitors of ribosome function in intact bacteria than in intact mammalian cells. Some antibiotics (e.g. cycloheximide) are specifically active against the larger ribosomes and are inactive against bacterial ribosomes. These have no clinical use since they are toxic to mammalian cells but they can provide useful biochemical tools for studying protein synthesis.

The structures of some of the most important antibiotics affecting ribosome function are shown in Fig. 6–2. With the possible exception of chloramphenicol, it can be seen that they are complex molecules and it is not surprising that their precise molecular interaction with the ribosomes is still uncertain. The sites of action of the three antibiotics to be discussed in this chapter are shown in Fig. 6–3. Many other antibiotics are also known to act upon the ribosome, but in many cases their detailed mechanism of action will not be understood until we know more about the process of protein synthesis itself.

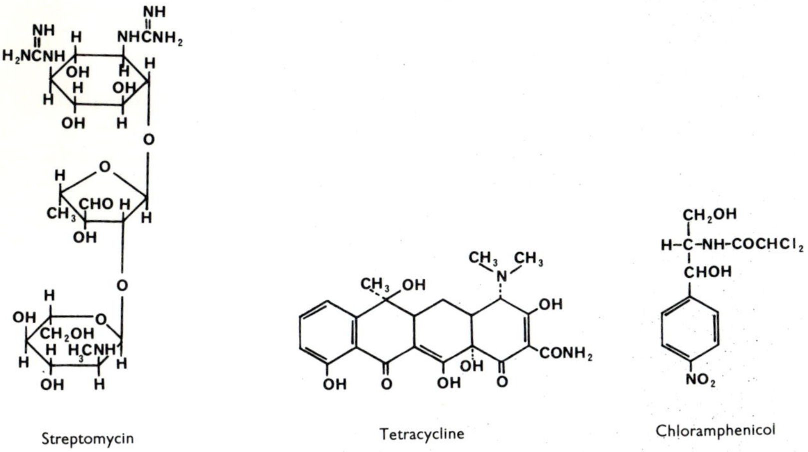

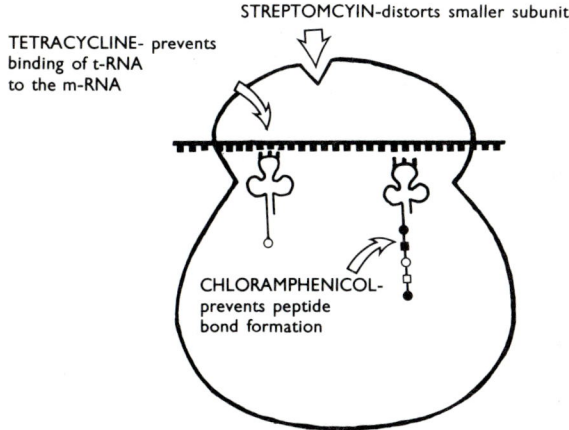

Fig. 6–2   Some antibiotics which interfere with ribosome function.

Fig. 6–3   Site of action of some antibiotics upon the ribosome.

6.2 Streptomycin

This is one of the antibiotics whose site of interaction with the ribosome has been studied in some detail. It is one of the few antibiotics effective against tuberculosis although its initial success in this field was limited by the emergence of resistance strains of tubercle bacilli. In addition it can have undesirable side-effects, e.g. irreversible deafness, in patients undergoing therapy.

Streptomycin reacts only with bacterial ribosomes, specifically with the smaller subunit of this organelle. The binding of the antibiotic distorts the ribosome enough to prevent normal interaction between the codon of mRNA and the anticodon of tRNA. This causes miscoding of the proteins produced leading to the production of nonsense-proteins, so effectively bringing normal protein synthesis to a halt. Bacterial resistance to streptomycin is due to the development of ribosomes with smaller subunits which do not bind the antibiotic. Conversely some organisms have been isolated which will only grow in the presence of streptomycin and in these it appears that the ribosome has to bind streptomycin molecules before it can function.

Addition of streptomycin to a culture of sensitive bacteria not only causes the inhibition of protein synthesis but also leakage of low molecular weight cytoplasmic components such as potassium ions; in addition it inhibits respiration. These observations suggest that streptomycin damages the cell membrane when it is taken up by the cells before it reaches its active site on the ribosome.

6.3 Tetracyclines

Tetracycline is the parent member of a group of useful broad spectrum antibiotics that are equally effective against Gram-positive and Gram-negative bacteria. They inhibit protein synthesis by ribosomes isolated from all kinds of cells but are far more active against bacteria than against mammalian cells. This may be because bacteria, unlike mammalian cells which are relatively impermeable to tetracyclines, actively take up tetracycline from solution and attain a high intracellular concentration of the antibiotic. It is thought that magnesium ions are involved in the uptake process. Tetracycline molecules form a complex with the magnesium ions that are present in the bacterial cell wall and on the surface of the cytoplasmic membrane. The complex is then transported across the membrane and the tetracycline molecules are released into the cytoplasm and act upon the ribosomes. The process results in an intracellular concentration of antibiotic up to 30 times the concentration outside the cells. How tetracycline interferes with ribosomal function is not entirely clear, but it appears to prevent the binding of the tRNA-amino acids to the acceptor site on the ribosome.

6.4 Chloramphenicol

This broad spectrum antibiotic was first isolated as a natural product but is now produced entirely by chemical means. It acts only upon bacterial ribosomes; one molecule binds to each of the larger subunits and this, by inhibition of peptide bond formation, is sufficient to prevent the incorporation of amino acids. Chloramphenicol is non-irritant and is used in the treatment of eye infections. It is also a useful agent against typhoid fever but it can exert undesirable side effects involving irreversible damage to bone marrow.

7 Metabolic Inhibitors

7.1 Introduction

The antibacterial compounds discussed in this chapter are all produced by purely chemical means. They are not therefore strictly speaking antibiotics, i.e. antimicrobial compounds produced by living cells. However, they are a group of compounds with considerable medical application, and are used alongside antibiotics particularly in the treatment of diseases caused by mycobacteria, such as leprosy and tuberculosis. Whereas many antibiotics act upon particular structures in the microbial cell, for example the cell wall, membrane or ribosomes, the metabolic inhibitors act upon enzymes involved in cellular intermediary metabolism.

7.2 Interference with folic acid

Folic acid (Fig. 7–1) and related compounds are co-enzymes concerned with the metabolism of groups containing one carbon atom: the transfer of methyl groups and the utilization of formate in the synthesis of amino

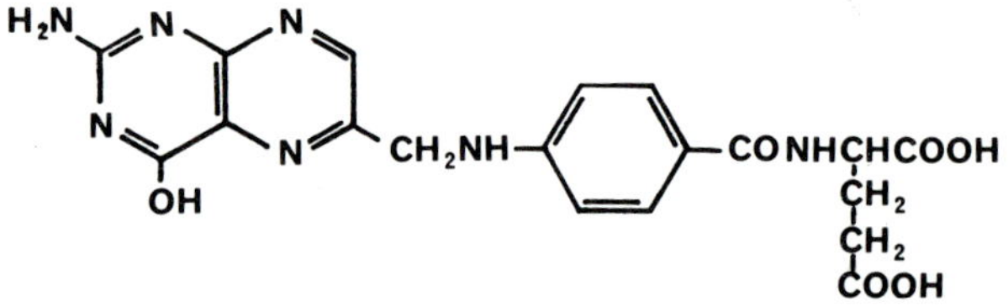

Fig. 7–1 The structure of the coenzyme folic acid.

acids and nucleotides. Mammalian cells cannot synthesize folic acid which is therefore an essential vitamin. In contrast, most disease-causing bacteria must synthesize folic acid and it is the inhibition of this synthetic pathway that forms the basis of the action of the inhibitors.

7.3 Sulphonamides

The sulphonamides were discovered in 1935 by Domagk as a result of testing numerous dyes for antibacterial activity. The compound prontosil (Fig. 7–2) was found to cure streptococcal infections in mice but it was without effect upon isolated bacteria. It soon became apparent that prontosil was broken down inside the mice to p-aminobenzenesulphonic acid amide (sulphanilamide) which was responsible for the antibacterial

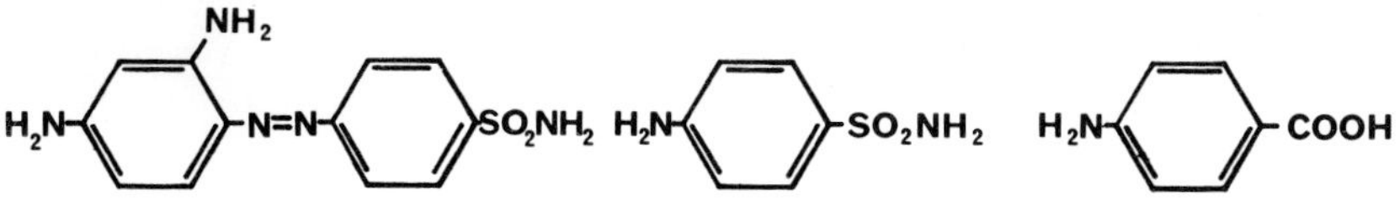

§ 7.4

Protonsil — Sulphanilamide — p-aminobenzoic acid (PABA)

Fig. 7–2 Prontosil is broken down in the body to give sulphanilamide which has a similar structure to p-aminobenzoic acid.

activity. When tested alone, sulphanilamide inhibited a wide range of bacteria. The key to the mechanism of action of sulphanilamide was the discovery that low concentrations of p-aminobenzoic acid (PABA, Fig. 7–2) cancelled out its inhibitory action. It was deduced that PABA is an essential metabolite for bacteria and that sulphanilamide competes with PABA as a substrate for enzymes involved in its metabolism. The basis of the competition was the marked structural similarity between the two molecules (Figs. 7–1 and 7–2). Sulphonamides exhibit a greater, or similar affinity for the enzyme tetrahydropteroic acid synthetase which is responsible for the incorporation of PABA into the folic acid molecule; i.e. sulphanilamides competitively inhibit the incorporation of PABA into folic acid. Some evidence suggests that sulphanilamide is sufficiently similar to PABA to be incorporated into folic acid itself giving a 'false' folic acid which is inactive. The precise mechanism of action is still not clear; it is certainly more complicated than originally supposed and many of its aspects remain unexplained.

However, this has not affected the development of the sulphonamide group of drugs. Soon after the discovery of sulphanilamide it was found that the molecule could be chemically modified without loss of activity. A large number of active sulphonamides have been produced, most of which are variations on the sulphanilamide molecule involving substitution of the sulphanilamide group. Sulphonamides have largely been superseded by antibiotics which exhibit greater potency; nevertheless, they are still used for treatment of certain urinary tract infections and in veterinary medicine.

7.4 Other compounds affecting folic acid metabolism

Two compounds which were prepared during the search for more active compounds related to the sulphonamides are diaminodiphenylsulphone (dapsone) and p-aminosalicylic acid (PAS) (Fig. 7–3). These were initially discarded since they had little action against common bacterial infections but they have since been found to be useful in treating infections caused by mycobacteria. Dapsone is one of the few drugs effective in the treatment of leprosy and PAS is used in the treatment of tuberculosis. Although they are thought to act in a similar way to the sulphonamides, it is not clear why they are particularly effective against mycobacteria.

Another group of drugs which affect folic acid metabolism is the pyrimidine derivatives. An example is trimethoprim (Fig. 7–3) which

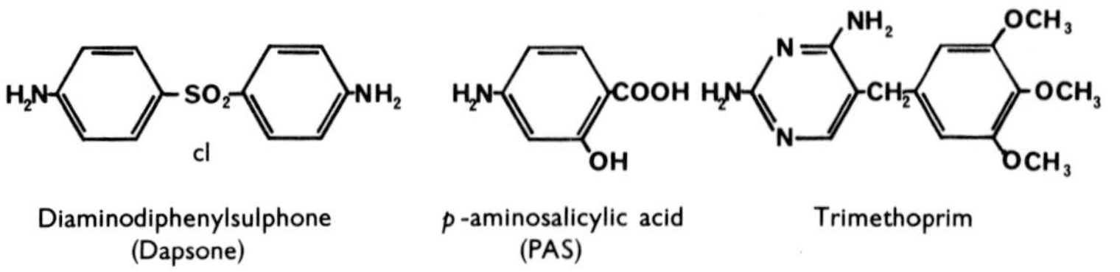

Diaminodiphenylsulphone (Dapsone)   p-aminosalicylic acid (PAS)   Trimethoprim

Fig. 7–3 Some other compounds which affect folic acid metabolism.

bears a structural resemblance to part of the folic acid molecule (Fig. 7–1). Trimethoprim has an inhibitory action against dihydrofolate reductase, one of the enzymes involved in folic acid metabolism. The enzyme from bacteria is far more sensitive than the corresponding enzyme found in mammalian cells, hence trimethoprim has a selective action against bacteria. It is usually used in combination with a sulphonamide; the resulting blockage of folic acid synthesis at two points is found to be particularly effective.

8 Prospects for Chemotherapy

8.1 Antibiotic resistance in microorganisms

In previous chapters we have described a veritable battery of antimicrobial agents capable of destroying any bacterial or fungal pathogen. Why, then, the continued search by scientists to discover new and better antibiotics; and if the antibiotics currently used appear to do their job adequately, why study their mechanism of action? The answer to both these questions comes from the inherent ability of microorganisms to adapt to hostile environments. The development of microbial strains resistant to antibiotics is unfortunately a common phenomenon. Some modern strains require much higher antibiotic concentrations than were necessary when the antibiotic was first used in therapy, while other strains have developed an absolute resistance. The penicillin dose presently used to control Gram-positive cells is several thousand times that used in the forties. For over thirty years penicillin was the drug of choice for the treatment of gonorrhoea but continued prolonged world-wide therapeutic use has led to the development of completely penicillin-resistant gonococci. There is an alternative agent, spectinomycin but if resistance to this antibiotic also develops the condition could easily become untreatable.

Many of the antimicrobial agents currently used were discovered twenty, thirty or forty years ago. The pace of antibiotic discovery has not abated, in fact it has increased, but few of the small number of new antibiotics that reach clinical trials show any advantage over existing antibiotics. It would not be alarmist to say that we are only a half-step ahead of the ability of many bacteria to overcome modern chemotherapy, with the possibility of a return to the dark days of the pre-antibiotic era. Hence the development of new and better antibiotics or the modification of existing antibiotics to increase their usefulness constitute very important aims of medicinal chemistry. Study of the mechanisms of antibiotic action not only provides a valuable insight into microbial biochemistry but is also a great help in the development of new and better drugs.

Antibiotic resistance may be divided into two categories: intrinsic resistance due to the physiology, biochemistry or morphology of the bacterial strain in question, and resistance acquired by microbial strains following exposure to antibiotics.

(a) *Intrinsic resistance*. Generally, Gram-negative bacteria are more resistant to antibiotics than Gram-positive bacteria, even though protoplasts prepared from Gram-negative and Gram-positive organisms often exhibit equal antibiotic sensitivity. Gram-negative organisms possess a lipid-rich layer enveloping the cell wall which is called the outer membrane. The layer is not present in Gram-positive bacteria (Fig. 8–1). The greater intrinsic resistance of Gram-negative organisms to antimicrobial agents may depend upon the non-specific permeability barrier presented by the outer membrane, preventing access of antibiotic molecules to their active site. If Gram-negative cells are treated with agents which selectivity remove part of the outer membrane they lose their intrinsic antibiotic resistance. For example treatment of *Escherichia coli* with ethylenediamine-tetraacetic acid (EDTA) removes the lipopolysaccharide (LPS) from the outer membrane and makes the cells more sensitive to a number of antibiotics.

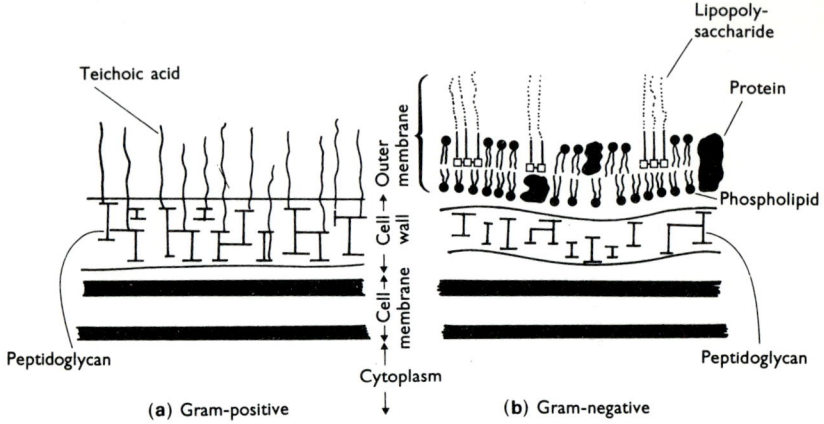

Fig. 8–1 Structure of the surface layers of Gram-positive and Gram-negative bacteria.

(b) Acquired resistance. Whenever a population of any organism is exposed to stress, natural selection will favour those individuals best fitted to survive. Antibiotics act as screening agents preventing the growth of sensitive cells and allowing the outgrowth of resistant mutants. The rate of mutation in bacteria is not particularly high (one gene mutation/10^5 to 10^7 cells/division) but the large numbers involved in bacterial populations make the chance of the development of bacteria resistant to a specific antibiotic high. Antibiotic resistance is not the result of specific mutations induced by the drug, indeed spontaneous mutation regularly occurs in the absence of the antibiotic against which resistance develops. Drug-resistant variants may be at a selective disadvantage in an antibiotic-free medium, since those changes in cell morphology/biochemistry conferring resistance may reduce their ability to compete with wild-type cells. Often if the selective pressure exerted by the anti-

biotic is removed, the resistant mutant will revert to the sensitive wild-type. Single mutants can confer a high degree of resistance (Fig. 8–2). Resistance increases of up to a thousand-fold over the parent strain have been reported after a single exposure to streptomycin or erythromycin. More commonly high levels of resistance develop as the consequence of a series of small increments (two to fivefold) in resistance (Fig. 8–2). With certain antibiotics no substantial degree of resistance has developed, e.g. polypeptide antibiotics, vancomycin and polyene antifungal antibiotics.

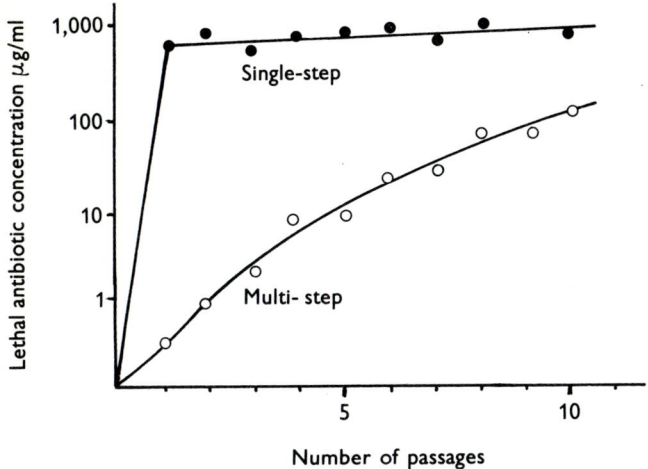

Fig. 8–2 Development of antibiotic-resistant mutants. Resistance can increase by a series of small steps after each exposure to the antibiotic (multi-step) or in one large step after the first exposure (single-step).

Detailed examination of the development of antibiotic resistance within microbial populations has shown that not all cases are the direct result of mutation. Resistant bacteria may transmit genetic information capable of conferring antibiotic resistance to sensitive strains by two mechanisms, transduction and conjugation. These processes are very complex and not fully understood. Only a broad outline of their mechanism will be presented.

In transduction DNA is carried from a donor to a recipient strain inside a bacterial virus or bacteriophage (phage). In this case the genes which render the bacterium resistant are not carried on the bacterial chromosome but on self-replicating extra-chromosomal elements, termed plasmids. They are small (approximately 1–2% of the bacterial chromosome) and like bacterial chromosomes are circular. Bacterial plasmids may become incorporated into the DNA of an invading phage. The incorporated plasmid replicates with the phage DNA within the bacterial cytoplasm, producing several hundred phage particles. These induce the bacterium to lyse releasing into the environment the new phage particles each of which carries a plasmid. If one of the new phage particles infects another bacterium, it might replicate, lyse the bacterium and release its progeny. Alternatively it might become temperate and not lyse the bacterial cell; in this case the genetic information of the bacterium, the phage and the plasmid exist and replicate side by side. All the progeny of such a bacterium would then be antibiotic-resistant through the expression of the plasmid genes. Transduction tends to occur within bacterial species reflecting the limited host ranges of most phages. Plasmid-carried resistance genes have been demonstrated in enterobacteria and pathogenic staphylococci for many commonly used antibiotics, including penicillin, chloramphenicol, tetracycline, kanamycin and erythromycin. Some of these genes, e.g. those conferring resistance to penicillin and erythromycin have been shown to be located on the same plasmid, while others are located on different plasmids.

Conjugation, i.e. 'sexual' contact between two bacteria, can also allow the transmission of antibiotic-resistance from resistant to susceptible bacteria. Many of the antibiotic resistance plasmids of enteric (gut) bacteria also contain genes involved in the promotion of their own transfer and are called R-factors. Their presence endows a bacterium with the ability to conjugate or 'mate' with a complementary bacterium, i.e. one without an R factor. Conjugation is achieved by the development of a cytoplasmic bridge, or pilus, between the donor (R^+) and recipient (R^-) strain, through which nucleic acid and the R-factor can pass. The recipient strain and its descendants are not only antibiotic-resistant, they also become able to conjugate and pass on the plasmid to other bacteria. R-factors can confer resistance to a variety of common antibiotics including chloramphenicol, tetracycline, streptomycin, penicillin, polymyxin and sulphonamides either individually or collectively. R-factors can be transmitted not only to members of the same species, but also to bacteria of other related enterobacteria including *Shigella, Salmonella, Proteus, Escherichia, Aerobacter, Serratia* and *Vibrio cholerae*. Thus R-factors can be transmitted from a pathogen such as *Shigella* to a non-pathogen, e.g. *E. coli*, and then back to other pathogens. *E. coli* is a universal inhabitant of the human gut and can therefore act as a reservoir of R-factors capable of rendering any invading pathogen resistant to antibiotics. Fortunately R-factor resistance can be lost spontaneously, often within months or even weeks of its acquisition; if this were not the case, widespread antibiotic resistance would make chemotherapy impracticable.

8.2 Biochemical nature of antibiotic-resistance

Microorganisms can achieve resistance to antibiotics through

§ 8.2   BIOCHEMICAL NATURE OF ANTIBIOTIC-RESISTANCE   57

biochemical, physiological and morphological modifications. Such changes include:

8.2.1   Modification of the bacterial site at which the antibiotic acts, making it insensitive to the drug, while still able to bring about its normal cellular function

This may involve changes in cellular morphology or biochemistry. The resistance of certain bacteria to streptomycin has been shown to result from specfic modifications to the structure of the bacterial ribosome which prevents the antibiotic binding to the organelle. Sulphonamides exhibit a greater affinity for the enzyme tetrahydropteroic acid synthetase than does its normal substrate PABA (see Chapter 7). Thus in the presence of equal amounts of sulphonamide and PABA bacterial folic acid biosynthesis is impaired. Sulphonamide resistance might arise by the development of a modified enzyme with a greater affinity for PABA than is possessed by the antimicrobial agent.

8.2.2   Enhancement of alternative metabolic pathways

The development of alternative metabolic routes to by-pass metabolic blocks caused by antimicrobial agents is not common but has been demonstrated in fungi resistant to antimycin A. This antibiotic inhibits the terminal stages of respiration. Antimycin-resistant fungi possess an alternative respiratory chain which is insensitive to the antibiotic. Pre-existing by-pass reactions may be enhanced to give bacteria resistance to antimicrobial agents. The pathways responsible for the salvage of purine and pyrimidine bases from nucleic acid catabolism, thereby allowing their re-utilization into new nucleic acids, may be used to circumvent the antibacterial activity of certain nucleic acid analogues.

8.2.3   Reduction in the physiological importance of the target site

Bacteria may overcome the antimicrobial action of sulphonamides by accumulating large amounts of PABA. Similarly certain penicillin-resistant mutants have been shown to possess reduced cross-linking in the peptidoglycan of their cell walls. In both cases resistance is conferred by reducing the importance to the cell of the site affected by the drug.

8.2.4   Preventing access of the antibiotic to its active site

The loss of cell permeability to the antimicrobial agent, with consequent inability of the drug to reach an internal concentration sufficient to damage the cell is often the result of specific changes in the cell envelope. Ampicillin-resistant mutants of *Salmonella typhimurium* and *Escherichia coli* have been shown to possess modified cell wall polysaccharides. The resistance of certain pneumococcal strains to streptomycin and erythromycin has also been explained by the development of an additional permeability barrier. Alternatively, in cases where the antibiotic is actively taken up by the microbial cell, drug uptake

58   BIOCHEMICAL NATURE OF ANTIBIOTIC-RESISTANCE   § 8.2

may be prevented by specific antagonism of the antibiotic transport mechanism. Tetracycline-sensitive Gram-negative and Gram-positive bacteria accumulate this antibiotic by an active process within the cell membrane which requires energy in the form of adenosine triphosphate (ATP). In resistant mutants exposure to tetracycline induces a block in this uptake mechanism. Similar changes in active uptake mechanisms have been demonstrated in mutants resistant to amino acid analogues. Such mutants show diminished ability to translocate the normal amino acid as well as its analogue.

8.2.5   Synthesis of enzymes capable of destroying the antibiotic

In clinical practice the most frequently encountered antibiotic-destroying organisms are Gram-positive bacteria which produce penicillinase, in particular certain strains of *Staphylococcus*. These strains owe their resistance almost entirely to the enzymic inactivation of the antibiotic, the individual bacterial cell wall assembly mechanism being as penicillin-sensitive as that of the parent strain. The enzyme originally called penicillinase breaks the β-lactam ring of the penicillin molecule (to give inactive penicilloic acid) and is better known as β-lactamase (Fig. 8-3). Most penicillins and most cephalosporins (which are chemically very similar to penicillins) are sensitive to staphylococcal β-lactamase, although some penicillins, notably methicillin and cloxacillin are resistant. Three distinct β-lactamase enzyme groups have been identified,

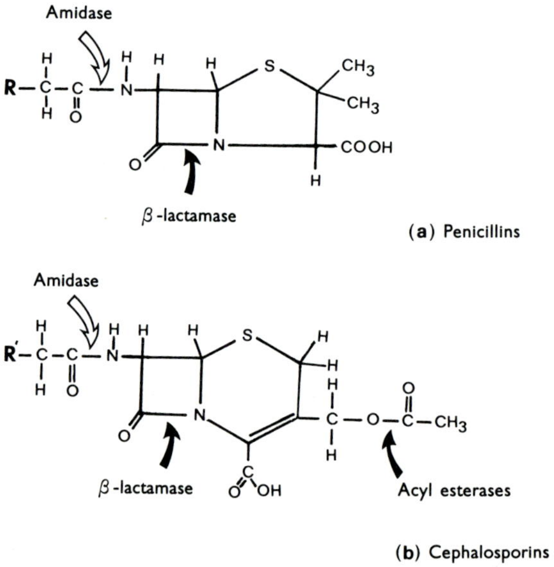

Fig. 8-3   Site of action of enzymes degrading penicillins and cephalosporins.

differing only in their affinity for the lactam ring and their immunological properties. A wide range of both Gram-positive and Gram-negative organisms exhibit β-lactamase activity although often the latter can be induced to produce the enzyme more easily and in greater amounts.

Two other enzymes, amidase and acyl esterase, are also capable of destroying the antimicrobial properties of penicillins and cephalosporins (Fig. 8–3) but although these enzymes are widely distributed in nature they play only a minor role in bacterial resistance.

Chloramphenicol-resistant Gram-negative and Gram-positive bacteria have been shown to be capable of an enzymic modification of chloramphenicol which leads to the loss of antimicrobial activity. The enzyme chloramphenicol acetyl transferase, which can be induced by chloramphenicol in *Staphylococcus aureus* but is constitutive in *E. coli*, brings about the acetylation of the antibiotic to give inactive mono- and di-acetyl derivatives. Enzymes capable of hydrolysing polymyxins and adenylating streptomycin have also been demonstrated in antibiotic resistant bacteria.

8.3 The spread and control of antibiotic-resistance

The development of antibiotic-resistant strains is a world-wide problem. The ease of modern travel allows the rapid spreading of such strains across the world. In any environment the more that antibiotics are used, the greater is the chance that the microbial population will develop resistance. Of the thousands of tons of antibiotics produced each year a considerable proportion is used non-clinically. Many scientists believe that the use of antibiotics, particularly those used in chemotherapy, should be restricted to cases where use has direct benefit to the patient. Non-clinical antibiotic use by the general public should be discouraged as it is, for example, in most of Europe and in N. America, but in many countries antibiotics may be easily bought by the general public. Modern animal husbandry relies on large numbers of animals being kept in close proximity, conditions which are ideal for the spread of disease. It has been found that dosing animal feed with antibiotics, or even antibiotic-containing by-products from drug manufacture, increases the health and weight-gain of animals. However, this sub-lethal dosing allied to poor sanitary conditions and the close proximity of other animals favours the development of resistant mutants and R-factors in the microbial flora of the animals. Because of the danger of resistant organisms spreading to humans, the practice has been forbidden in the UK. It has also been suggested that certain clinical antibiotics would make excellent food preservatives, but such a use would be irresponsible.

The clinical use of antibiotics has led to some special problems concerned with bacterial resistance. Because of their constant contact with antibiotics and patients undergoing chemotherapy, hospital personnel often possess a highly resistant microbial flora which could be passed on to other patients. Cross-infection between patients is also common. In an attempt to reduce resistance levels many hospitals periodically change the antibiotics they generally use. The 'wonder drug' image of many antibiotics has in the past led to their being prescribed for the treatment of common viral diseases, e.g. flu and the common cold, against which they have no effect. Treatment with sub-lethal antibiotic levels and unnecessary therapy also serve to increase the general degree of resistance within the microbial population. If bacterial resistance develops during antibiotic therapy, the treatment must be immediately changed to an alternative drug, often before further antibiotic sensitivity testing can be carried out. In such cases, and in cases of infection by an organism of unknown sensitivity multiple antibiotic thereapy is used. The use of a combination of lethal concentrations of two unrelated drugs restricts the development of resistant strains. If the mutation rate required to produce bacteria insensitive to each of two drugs used alone is 10^{-6}/bacterium, then the mutation rate necessary for the development of a strain resistant to both agents used together is the product, not the sum, of the individual frequences, i.e. 10^{-12}/bacterium (not 2×10^{-6}).

8.4 Development of new antimicrobial agents

Since the original observations of antibiotics, several hundred thousand synthetic and natural products have been screened for antibiotic activity. Of the many thousands found to inhibit bacteria, less than 5% have ever been used clinically, and less than 0.1% would be considered suitable for general use. The search for new antibiotics continues using empirical methods similar to those described in Chapter 2. This approach has many disadvantages. It is expensive and requires many time-consuming tests. Cost and time limitations restrict the search to certain genera of antibiotic-producing organisms and the number of strains and culture conditions against which a potential new antibiotic can be tested. Another inadequacy of antibiotic screening is that it fails to detect agents which, while showing little or no effects upon bacteria when used alone, might make a resistant organism more sensitive to another antimicrobial agent or to the natural defence mechanism of the patient. Hence the full potency of many existing antibiotics may never be known.

Many scientists hope that one day it will be possible, through an understanding of metabolite structure and the ways in which metabolites bind to enzymes, co-enzymes or other products, to design antimicrobial agents on a rational basis. Similarly by introducing substituents of known physical and chemical properties to such molecules, it should be possible to design and synthesize potent new antimicrobial agents with wider antimicrobial spectra, increased rates of microbial penetration or affinities for the target sites, or with diminished toxic side effects. It would

§ 8.4 DEVELOPMENT OF NEW ANTIMICROBIAL AGENTS

be pleasing to report that such an approach had been successful; however, despite many attempts to produce such agents, particularly over the last ten years, very few useful agents have been produced and these in no way compare in antimicrobial activity with the large number of agents discovered by the screening of microbial secondary metabolites. Scientists, as yet, do not know enough about microbial physiology or the subtle changes elicited by antibiotics to mimic their action.

One approach which has had limited success has been the chemical modification of existing antibiotics. This use of the nucleus of an existing antibiotic molecule overcomes one of the main difficulties of designing novel antibiotics i.e. the production of structural analogues capable of covalent bond formation with a target site. Some of the successes are only marginal while others, including the development of synthetic derivatives of rifamycin (see Chapter 5) and more notably the semi-synthetic penicillins and cephalosporins, have distinct advantages over the parent antibiotic. Penicillin V and G synthesis can be induced by changing the culture conditions of the producing fungi. Further derivatives can be obtained if the natural penicillin is treated with amidase (Fig. 8–3) to produce 6-aminopenicillanic acid. This compound can be chemically modified to give a series of semi-synthetic penicillins (Fig. 8–4) which

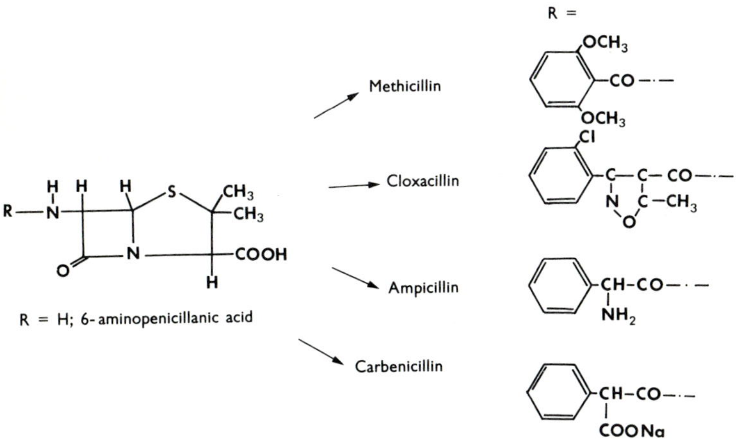

Fig. 8–4 Chemical modification of penicillins.

possess several advantageous properties (Table 4) over natural penicillins, including increased antimicrobial spectrum, activity, resistance to β-lactamase and acid resistance. This last property allows the antibiotic to resist stomach acid and therefore permits oral dosing.

Table 4 Properties of some natural and semi-synthetic penicillins (++ very effective; + effective; − ineffective)

	Antimicrobial spectrum		Oral dosing	
	Gram+ve bacilli	Gram−ve bacilli	β-Lactamase resistance	Acid resistance
Natural				
Penicillin G	++	−	−	−
Penicillin V	++	−	−	+
Semi-synthetic				
Methicillin	+	−	+	−
Cloxacillin	++	−	++	+
Ampicillin	++	+	−	+
Carbenicillin	++	++	−	−

8.5 Antibiotics as tools in scientific research

Many of the major advances of biochemistry, physiology and molecular biology of the last twenty years have been made with the assistance of antibiotics. They have proved potent agents, not only in probing the cellular activity of bacterial cells but also in determining the mechanisms controlling the function and development of mammalian cells. By using different antibiotics cellular metabolism can be blocked at specific sites, and by examining the consequences of such blockages the nature of the pathway and the factors controlling it can be studied.

Our knowledge of the structure and biosynthesis of the bacterial cell wall is due in large part to the existence of penicillin. Chloramphenicol has contributed to our knowledge of enzyme induction, the origin and direction of chromosome replication, RNA and protein synthesis, and ribosome function. Ionophores, such as valinomycin, were instrumental in developing modern concepts of membrane transport. One antibiotic which is too toxic to be used in chemotherapy, puromycin, has proved invaluable in understanding the peptide-forming steps of protein synthesis. Inhibitors of nucleic acid synthesis, e.g. actinomycin and mitomycin have been used successfully to study the molecular basis of heredity.

8.6 The future

Chemotherapy has changed dramatically over the last twenty years and it is very difficult to predict even the immediate future. Antibacterial therapy will remain the main role of antibiotics in medicine provided that the spread of antibiotic resistance can be halted. The careful and proper use of antimicrobial agents can, and it is hoped will, dramatically reduce the degree of antibiotic resistance in microbial populations. New antibiotics will be discovered and existing ones modified to increase their antimicrobial activity while reducing their toxic side effects. Many antibiotics kill a wide range of bacteria and often disturb the hosts' natural flora while destroying an invading pathogen. Disturbance of the natural flora may leave the host open to new infection. The development of agents lethal only to specific pathogens would be a useful addition to chemotherapy.

The major success of antibiotics in medicine has been their ability to control bacterial, and to a lesser extent mycotic, pathogens. Antibiotics have found little application in defeating protozoan infections (e.g. malaria, sleeping sickness) or viral infections (e.g. polio, rabies, influenza). The discovery of agents to control these conditions would be as great a step forward as the initial discovery of antibiotics. Certain antibiotics have possible application as anti-tumour agents; gramicidin and puromycin possess carcinolytic properties; L-azaserine, sarkomycin and nucleotide analogues have been found to inhibit tumours in rodents and other experimental animals. The successful use of antibiotics in anti-cancer therapy is still a long way off but its very possibility opens a whole new field of antibiotic research.

Further Reading

ALBERT, A. (1975). *The Selectivity of Drugs*, Chapman and Hall, London.
BENVENISTE, R. (1973). Mechanisms of Antibiotic Resistance. *Annu. Rev. Biochem.*, **42**, 471.
BUCHER, T. H. and SIES, H. (1949). *Inhibitors: Tools in Cell Research*. Springer Verlag, Berlin.
FRANKLIN, T. J. and SNOW, G. A. (1975). *Biochemistry of Antimicrobial Action*. Chapman and Hall, London.
GALE, E. F., CUNDLIFFE, E., REYNOLDS, P. E., RICHMOND, M. H. and WARING, M. J. (1972). *The Molecular Basis of Antibiotic Action*. John Wiley, London.
GARROD, L. P., LAMBERT, H. P. and O'GRADY, F. (1973). *Antibiotic and Chemotherapy*, 4th ed. Churchill Livingstone, Edinburgh.
HAYES, W. (1968). *The Genetics of Bacteria and their Viruses*. Blackwell Scientific Publications, Oxford.
HUGO, W. B. (1971). *Inhibition and Destruction of the Microbial Cell*. Academic Press, London.
LEIVE, L. (1973). *Bacterial Membranes and Walls*. M. Dekker, New York.
MUNOZ, E., GARCIA-FERRANDIZ, F. and VAZQUEZ, D. (1972). *Molecular Mechanisms of Antibiotic action on Protein Biosynthesis and Membranes*. Elsevier, Amsterdam.
PRATT, W. B. (1973). *Fundamentals of Chemotherapy*. Oxford University Press.
SALTON, M. R. J. and TOMASZ, A. (1974). Mode of action of Antibiotics on Microbial Walls and Membranes. *Ann. N. Y. Acad. Sci.*, **235**.
WATSON, J. D. (1965). *Molecular Biology of the Gene*. Benjamin, New York.